Abdurahamon Ergashev
Obid Yunusov
Elshod Ulugmurodov

Peculiarities of manufacturing of three-electrode temperature sensors

Abdurahamon Ergashev
Obid Yunusov
Elshod Ulugmurodov

Peculiarities of manufacturing of three-electrode temperature sensors

Three-electrode silicon structures for temperature sensors

This book is a translation from the original published under ISBN 978-620-2-39640-0.

Publisher:
Sciencia Scripts
is a trademark of
Dodo Books Indian Ocean Ltd. and OmniScriptum S.R.L publishing group

120 High Road, East Finchley, London, N2 9ED, United Kingdom
Str. Armeneasca 28/1, office 1, Chisinau MD-2012, Republic of Moldova, Europe
Printed at: see last page
ISBN: 978-620-7-95191-8

Contents

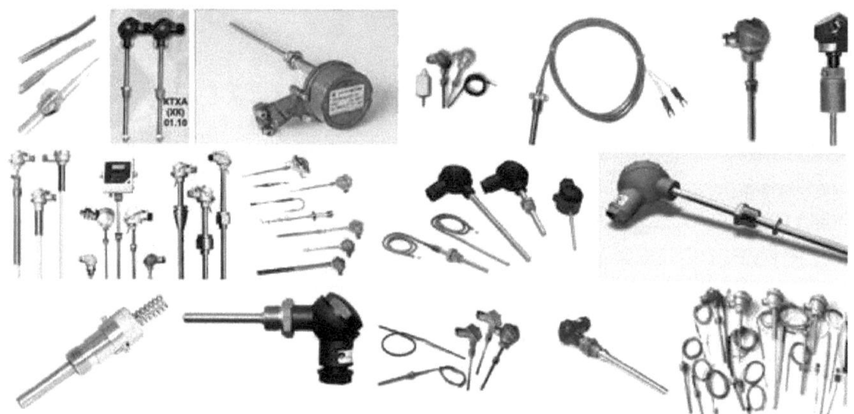

In the monograph, *three-electrode silicon structures for temperature sensors are presented. For the study, 2 types of silicon p-n structures were fabricated, consisting of an n-type epitaxial layer doped with phosphorus in type IU samples in type II samples, which is grown on a p-type silicon substrate oriented in-plane and doped with boron.*

The research consists in the development of a two-level control system of the dosing process control in fabric bleaching. In this case, at the lower level the issue of maintaining technological parameters of a certain value at an acceptable level is solved, and the coordination of actions of technological units is carried out at a higher level. An effective algorithm for calculating a compensating device for evaluating the effect of changing the transfer function coefficients of digital regulators is proposed. It is proposed to use SCADA-system to control this technological process. The algorithm of modelling and synthesis of the control system with delay on the basis of dynamic graphical models is proposed, which allows to take into account the influence of delay on the object properties and compensate it.

INTRODUCTION

The modern stage of development of electronics and computer technology has led to the prerequisite for wide automation of the most diverse processes in industry, in scientific research and in everyday life. However, the practical realisation of this prerequisite is largely determined by the capabilities of devices designed to obtain information about the regulated parameter or process. Such devices are called sensors in technology [1-5]. Of course, the application of sensors is not limited only to automated systems, as they can also fulfil the functions of elements of simple measuring systems.

Currently, electronic thermometers for temperature measurement use thermodiodes, thermotransistors, resistors and integrated thermosensors [1,2]. Among other things, Uzbekistan is working on the development of temperature sensors in the form of silicon thermoresistors with a compensated base area: using diffusion technology by doping with transition metal impurities (manganese, nickel, etc.) [3], and by thermal radiation doping of silicon [4]. In addition, based on integrated circuits on complementary transistors consisting of several dozens of transistors, a number of thermal sensors capable of monitoring the temperature of various objects have been proposed [5,6].

A common disadvantage of known semiconductor sensors is their low accuracy of temperature measurement. The accuracy of temperature measurement in semiconductor sensors is determined by a combination of many factors, most of which is a consequence of the strong dependence of temperature sensitivity of these structures on technological and material parameters, which due to non-ideal reproducibility of technological production of semiconductor devices have a fairly large variation. In addition, the presence of temperature dependence of such parameters as non-ideality coefficient or series resistance of the structure leads to a significant decrease in measurement accuracy.

The analysis of the temperature sensor market also showed that a significant nomenclature among them is occupied by measuring and converting devices for the temperature range from -100 to +200 °C. For sensors with combined or integrated sensing elements produced by the industry, the range of measured temperatures is determined not only by the capabilities of the sensing element, but also by the temperature range, in which the operability of the measuring information processing unit combined with the sensing element is ensured. The best foreign samples of semiconductor integrated temperature sensors [6-15] today can provide temperature measurements in the range from 0 °C to +125 with an error of ±2.0 °C. However, there is a need in science and industry to monitor and measure higher and lower temperatures [16, 17]. Thus, for example, in the steam heating system of urban and rural areas objects require highly accurate (not worse than ±1 °C) water vapour temperature measurements of the order of (100-150) °C. In flying space and aviation vehicles, on the contrary, it is necessary to control low temperatures, of the order of (-75-10) °C. Depending on the type of object on flying aircraft the number of temperature sensors can reach up to 1000. Therefore, the creation of a cheap and reliable integrated sensor for the range of measured temperatures from -65 to +175 ° C, differing from similar foreign samples with improved operational and metrological characteristics and suitable for mass production by domestic electronic industry is an urgent scientific and technical problem.

The global market is currently selling billions of dollars worth of temperature sensors per year. And this is understandable: a nuclear power plant has about 1500 temperature measurement points for controlling automation and safety processes, and a large chemical industry enterprise has more than 20 thousand such points. The temperature range from -65 °C to +175 °C is practically not mastered for measurement by silicon integrated sensors, therefore in this work as an object of research the temperature sensor made in the form of silicon diode structure with built-in sensing element on the basis of p-p structures is chosen.

CHAPTER 1

. ANALYSING THE CURRENT STATE OF DEVELOPMENT IN THE FIELD OF TEMPERATURE SENSORS

1.1 Methods and means of temperature measurement

Temperature is a physical quantity directly proportional to the average kinetic energy of particles of matter (molecules or atoms) and characterising the state of thermodynamic equilibrium of a macroscopic system. In an isolated system that is not in equilibrium, over time the energy transfer from the more heated parts of the system to the less heated ones leads to temperature equalisation in the whole system [8,9].

Kelvin showed [10] that if one assigns a certain number of degrees to any value of the average kinetic energy of particles (the so-called "*reference point*" (As a reference point, the 10th General Conference on Weights and Measures in 1954 adopted the triple point of water, attributing to it the exact value of temperature 273.16 K by definition.), it is sufficient to construct a linear infinite temperature scale from absolute zero. The temperature is the same for all parts of an isolated system in thermodynamic equilibrium. If the isolated system is not in equilibrium, then over time the transfer of energy (heat transfer) from more heated parts of the system to less heated parts leads to equalisation of temperature in the whole system (the first postulate or the zero beginning of thermodynamics).

Temperature determines the distribution of particles forming a system by energy levels (Boltzmann statistics [10]) and the distribution of particles by velocities (Maxwell distribution [10]), the degree of ionisation of matter (Saha formula [10]), the property of equilibrium electromagnetic radiation of bodies and the spectral density of radiation (Stefan-Boltzmann radiation law), etc. The temperature included as a parameter in the Boltzmann distribution is often called the excitation temperature, in the Maxwell distribution - the kinetic capacitance temperature, in the Saha formula - the ionisation temperature, in the Stefan-Boltzmann law - the radiation temperature. Since for a system in thermodynamic equilibrium all these parameters are equal to each other, they are simply called the temperature of the system. In the kinetic theory of gases, the temperature is quantitatively defined in such a way that the average kinetic energy of translational motion of a particle (possessing three degrees of freedom) is equal to the temperature [11,12].

where m, V - mean mass and velocity of particles; T - temperature; k - Boltzmann constant equal to $1.38 \cdot 10^{-23}$ J/grade.

$$\frac{mV^2}{2} = \frac{3}{2}kT \qquad\qquad (1.1)$$

$$T = \frac{mV^2}{3k} \qquad\qquad (1.2)$$

In general, temperature is defined as the derivative of the energy of the body as a whole by its entropy. Such a temperature is always positive (because kinetic energy is positive), it is called absolute temperature or temperature on the thermodynamic temperature scale.

The inventor of the first thermometer-thermoscope was the famous Italian scientist Galileo Galilei (1597) [13].

Fig. 1.1 shows Galileo's device: a narrow glass tube 2, lowered into a vessel with water 3, was soldered to a glass ball 1. The ball was heated in the hands, water was lowered down the tube and set at some level above the water in the vessel 3. As the ball cooled, the air in the tube was compressed and the water level in the tube rose. Thus, by the position of the water level in tube 2 it was possible to judge qualitatively about the change in the temperature of the air in it. For convenience of observation, a scale with arbitrarily drawn divisions on it was attached to tube 2.

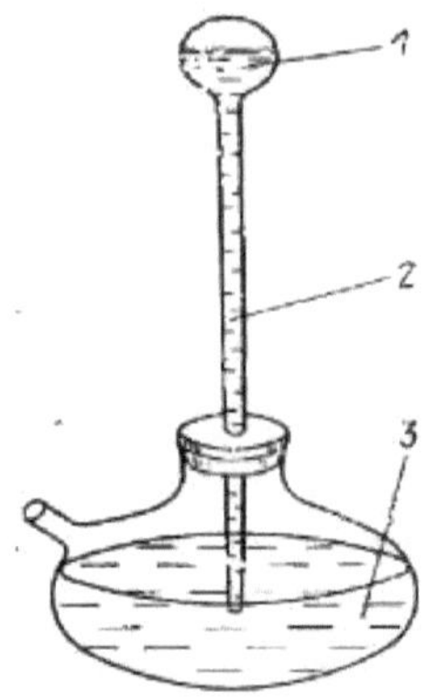

Figure 1.1. Galileo's thermoscope.

Modern thermometry has a variety of measurement methods, each of which is specific and not universal. Methods of temperature measurement are divided into *contact* (the measuring instrument is in direct *contact* with the controlled object) and *non-contact* (convenient for measuring high temperatures). The most accessible, accurate and reliable are the contact methods realised with the help of *thermometers* [14].

Gas thermometers utilise the direct proportional relationship between the pressure of an ideal gas and its absolute temperature at constant volume (Charles' law). They are used at low pressures and sufficiently high temperatures as reference thermometers, and other thermometers are calibrated and checked by them (Fig. 1.2). The disadvantage of gas thermometers is that they are bulky and inconvenient to use [8,14].

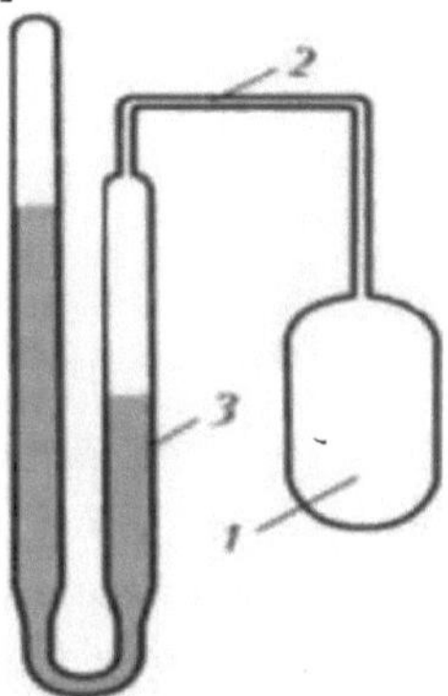

Fig. 1.2. Gas thermometer: 1-balloon (filled with gas); 2-coupling tube; 3- pressure measuring device.

Liquid thermometers are convenient to handle, but the range of their operation is limited to the crystallisation and boiling temperatures of liquids [6,14]. For example, we can say that mercury thermometer and alcohol thermometer are known and widely used, their principle of operation is based on the response of the liquid to the effect of temperature and is based on the change in the volume of the liquid with the change in temperature (when the liquid is heated, it expands and the alcohol column rises up, and when the temperature drops, it drops down). In medicine, the basic thermometer is the mercury thermometer, which, however, has an obvious disadvantage. Given the decisions of the Minamata Conference held in Japan under the auspices of the United Nations to ban the use of mercury thermometers for civilian purposes from 2021, there is an additional incentive to use mercury-free thermometers (Fig. 1.3). In this regard, mercury-free semiconductor thermometers and thermometers are promising, which, moreover, can be conveniently integrated into various micro-technical devices of miniaturised dimensions [7].

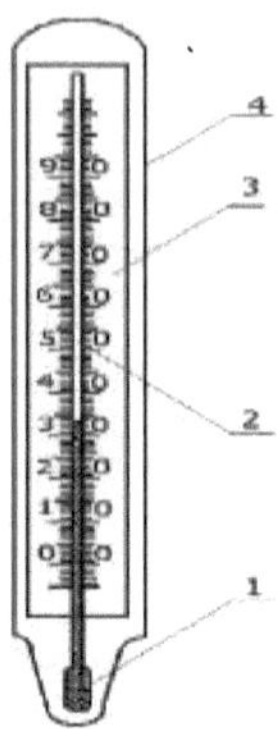

Figure 1.3. Glass liquid thermometer.

1-tank, 2-capillary, 3-scale, 4-glass protective shell.

A thermocouple is formed, as is known, by connecting two dissimilar conductors (Fig. 1.4), the ends of which are soldered together to form a ball [8]. The principle of operation of a thermocouple is based on the Seebeck effect, when in an electric circuit made up of different conductors, a thermal EMF occurs if the contact areas are at different temperatures[17,18].

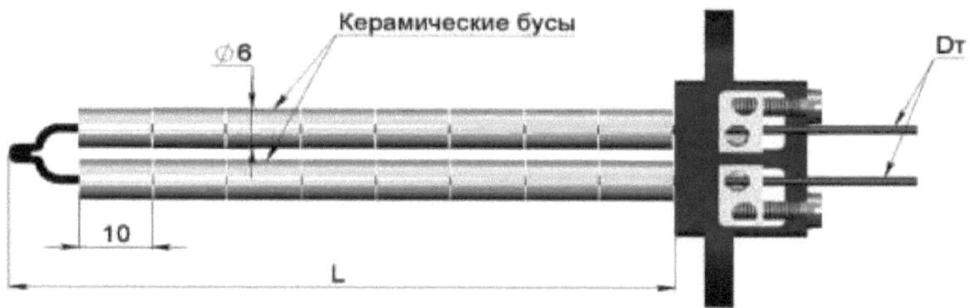

Figure 1.4: A conventional thermocouple.

Thermoresistive sensors, which are based on the change in resistance of a conductor when its temperature changes (Fig. 1.5). Resistance thermometers made of metals can measure temperature in a wide range, for example, a wire spiral made of platinum allows measurements from -258°C to 900°C. Their advantages are miniaturisation, good linearity of characteristics, high stability. The disadvantages include low sensitivity (resistance of metals increases with the

temperature increase at a rate of 0.4-0.6 % /K) and relatively large inertia [8,19].

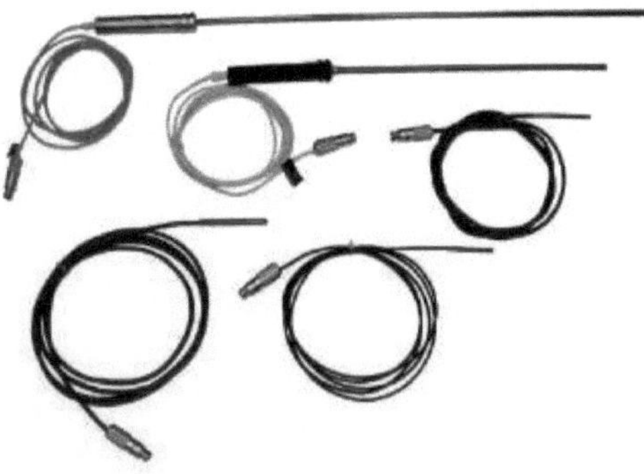

Fig. 1.5. Resistance thermometer.

Semiconductor sensors work on the principle of changing the characteristics of the p-n junction under the influence of temperature [14]. In particular, the WAC behaviour at direct bias of the p-n junction allows the p-n junction to be used as a semiconductor thermometer [20]. Therefore, diodes and transistors can be used to measure temperature (Fig.1.6). The advantages of such a solution are cheapness and linearity of characteristics over the entire measurement range[21].

Figure 1.6. Semiconductor temperature sensor.
In Acoustic Temperature Sensors, the principle of operation is different speed of sound in the medium at different temperature [22]. Knowing the initial data, it is possible to calculate temperature changes by the speed of sound wave travelling in the substance (Fig. 1.7). This is a non-contact method that allows measuring temperature in closed cavities, as well as in a medium inaccessible to direct

measurements. Such sensors are used in medicine and industry - where penetration to the substance to be measured is impossible [23].

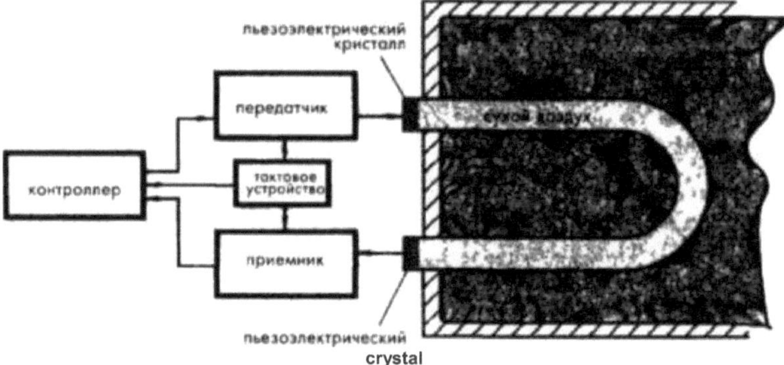

Fig.1.7. Acoustic thermometer with ultrasonic detector.

The non-contact type of temperature sensors (Pyrometers) reads the radiation that comes from heated bodies [23,24]. This type of devices allows measuring the temperature remotely, without approaching the medium in which the measurements are made (Fig. 1.8). This allows working with large temperatures and highly heated objects without dangerous proximity. All pyrometers according to the principle of operation are divided into interferometric, fluorescent and sensors based on solutions that change colour depending on the temperature [25,26].

Main characteristics of the DT-8818H pyrometer:
- measuring range -50 to +550 °C
- 16:1 optical ratio
- resolution of 0.1 °C
- accuracy ±5°C (-50 to -20), ±1.5% (-20 to 199), ±2.0% (200 to 550)
- integrated laser designator
- spectral range 8.14 μm
- adjustment of the emission factor from 0.1 to 1

Piezoelectric temperature sensors operate using a quartz piezoresonator. The whole essence of operation is in the direct piezo effect, i.e., in the change of linear dimensions of the piezo element under the influence of electric current [27]. At alternate supply of different-phase 11

The piezoresonator oscillates at a certain frequency, and the frequency of its oscillation depends on the temperature. Knowing this relationship, it is easy to convert the frequency of oscillation of the resonator into temperature. Due to the wide range of measurements and high accuracy, such sensors are mainly used in

research and experiments where high reliability and durability are required [28].
Each type of thermometer can be used in a specific temperature *range* and has
its own *advantages* and *disadvantages*.

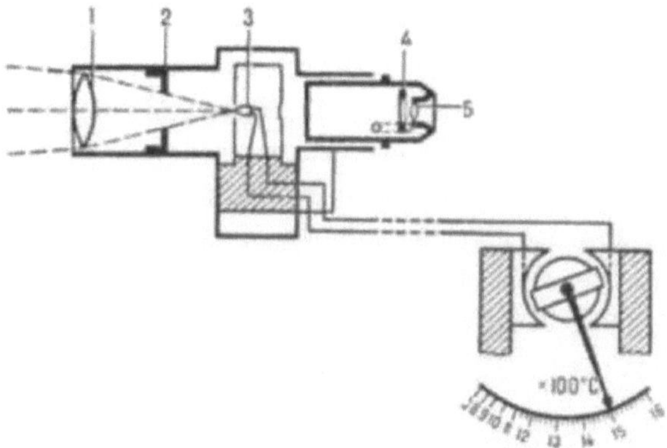

Fig.1.8. Full radiation pyrometer: 1-lens; 2-diaphragm; 3-receiver of radiation
(transducer); 4-ocular; 5-light filter.

1.2 Temperature scales.

On the Celsius scale, the temperature of the triple point of water is
approximately 0.008 C [10,12] and, therefore, the freezing point of water at a
pressure of 1 atm. is very close to 0 C. The boiling point of water, originally
chosen by Celsius as a second reference point with a value by definition equal to
100 C, has lost its status as one of the reference points. According to modern
estimates, the boiling point of water at normal atmospheric pressure in the
thermodynamic Celsius scale is about 99.975 C. The Celsius scale is very
convenient from a practical point of view, since water and its states are common
and essential to life on Earth. Zero on this scale is a special point for
meteorology because it is associated with the freezing of atmospheric water. The
scale was proposed by Anders Celsius in 1742.

Fahrenheit scale. [10,29] In England and especially in the United States, the
Fahrenheit scale is used. Zero degrees Celsius is 32 degrees Fahrenheit and 100
degrees Celsius is 212 degrees Fahrenheit. The current definition of the
Fahrenheit scale is as follows: it is a temperature scale with 1 degree (1 F) equal
to 1/180 of the difference between the boiling point of water and the melting
point of ice at atmospheric pressure, and the melting point of ice has a
temperature of +32°F. The temperature on the Fahrenheit scale is related to the
temperature on the Celsius scale (t °C) by the relationship t °C = 5/9 (t °F - 32),

$$t°F = 9/5 * t°C + 32 \qquad (1.3)$$

Proposed by G. Fahrenheit in 1724.

In physics, the most used is the thermodynamic (absolute) temperature scale (Kelvin scale) based on the second principle of thermodynamics. It has one reference point - the triple point of water [10,29], which is assigned the value $T_{TP} = 273.16K$. Temperatures on the Kelvin scale are counted from absolute zero temperatures, at which there is no any thermal motion of molecules. The Kelvin scale became the basis for the international standard of modern thermometry. The advantages of the scale are independence from the properties of the thermometric substance and high accuracy of reproduction of the triple point T_{TP}. Relationships between temperatures expressed on the Celsius scale and on the absolute thermodynamic scale:

$$T = t^o C + 273,15K \qquad (1.4)$$

1.3 Temperature sensors using diode and transistor structures as sensing elements.

1.3.1. Diode temperature sensors.

The use of silicon diode structures as primary temperature transducers allows to improve significantly the linearity of the temperature characteristic of the semiconductor temperature sensor in *comparison* with a silicon resistor. Indeed, if a constant current Inp is passed through the diode in the forward *direction*, its relation with the forward voltage Unp at the *p-n* junction of the diode is given by the known equation [30]:

$$I_{np} = I_{o6p} * (e^{\left(\frac{q*U_{np}}{kT}\right)} - 1) \qquad (1.5)$$

where k is the Boltzmann constant, q is the electron charge, T is the temperature in Kelvin, I_{o6p} is the reverse current through the p-n junction.

$$U_{np} \cong (\frac{kT}{q}) * \ln(\frac{I_{np}}{I_{o6p}}) \qquad (1.6)$$

Similarly :

The parameter determining the temperature dependence of the voltage at the p-n junction of the diode in equation (1.6) is the current I_{O6P} . Based on the general theory of the p-n junction, it is shown that in a limited temperature range (for silicon diodes from -50°C to +120°C) the direct voltage drop U_{np} is linearly U_{np} dependent on temperature with a temperature coefficient

$(TKH = \dfrac{dU_{np}}{dT}) \approx 1,5$мВ/° C

depending on the diode type and on the current density

junction. Moreover, as the forward current through the p-n junction increases, *TKH* decreases, and as the current *IP* decreases, the temperature range in which *TKH* can be considered constant decreases due to the influence of the current *ъbr*. Therefore, when diodes are used as sensing elements in integrated temperature sensors, the efforts of the developers were aimed at reducing the influence of the reverse current *Iобр* on the temperature dependence of the voltage *Uпp* and improving the operational characteristics of the sensors. However, the most promising, from the point of view of using diode sensing elements as primary transducers and mass production of semiconductor integrated temperature sensors, was p-n junction transistors [30, 31]. Two alternately different currents Inpi and 1_n p2 in the forward direction with respect to the emitter-base p-n junction are passed through the transistor V in diode inclusion (Fig. 1.9), as shown in Fig. 1.10, and a high level of injection is ensured for these currents, i.e., Inpi and *I p2* in the forward direction with respect to the emitter-base p-n junction.

$I_{пp}2 > I_{пp}1 >> I_{обp}$ и $U_{пp}1 >> kT/q$, , as shown in Fig. 1.11.

In this case (Fig. 1.9), equation (1.6) can be rewritten for two currents (*Inpi* and *Inp2*) and for two temperatures (T1 T2) as follows:

$$U_{пp1} = (\frac{kT_1}{q}) * \ln(\frac{I_{пp1}}{I_{обp}}) \qquad (1.7)$$

Accordingly, at current through the diode *IП= IПp2* and at temperature *T=Ti*, equation (1.6)

$$U_{np2} = (\frac{kT_2}{q}) * \ln(\frac{I_{np2}}{I_{обp}}) \qquad (1.8)$$

will be rewritten as:

A change in the forward current through the diode from the value of *Inpi* to the value of *IПp2* will cause the voltage across the diode (Fig. 1.9) to change by an amount:

$$\Delta U_{np.T1} = U_{np2} - U_{np1} = (\frac{kT_1}{q}) * \ln(\frac{I_{np2}}{I_{обp}}) \qquad (1.9)$$

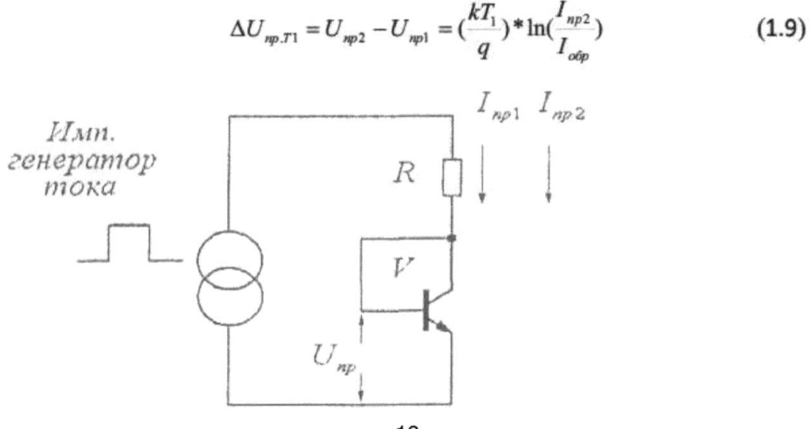

Figure 1.9. Transistor V in diode inclusion as a sensing transistor elements.

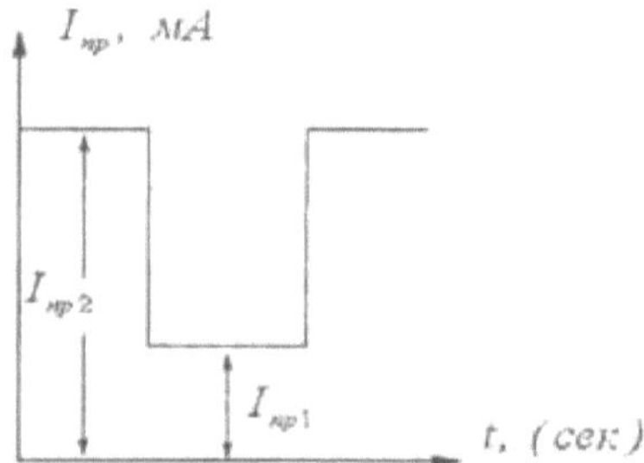

Fig. 1.10. Diagram of currents $I_{\Pi p1}$ and $I_{\Pi>2}$ flowing through the emitter-base diode.

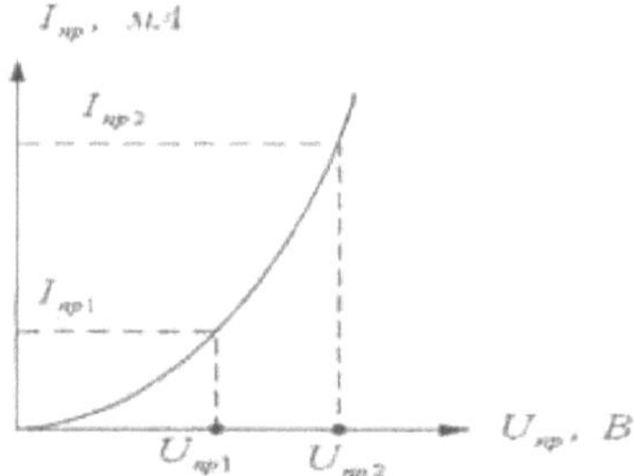

Fig. 1.11. Voltampere characteristic of *the* diode and voltage values corresponding to direct currents I_n p1 and I_{np2} flowing through the diode.

$$\Delta U_{np.T2} = U_{np2} - U_{np1} = (\frac{kT_2}{q}) * \ln(\frac{I_{np2}}{I_{np1}}) \qquad (1.10)$$

If we subtract equation (1.8) from equation (1.10), we obtain [40]:

$$\Delta U_{np} = U_{npT2} - U_{npT1} = (T_2 - T_1)\frac{k}{q} * \ln(\frac{I_{np2}}{I_{o6p}}) \qquad (1.11)$$

Equation (1.11) differs from equation (1.6) in that it does not contain such a diode parameter as I_{o6p} and that the increment of the direct voltage drop across the diode is directly proportional to the change in ambient temperature.

The pulsed electric mode can be easily replaced by the constant current electric mode [31], and the voltage ΔU_{np} is taken as the difference between the diode bases (Fig. 1.12).

For the circuit (Fig. 1.12), equation (1.12) can be rewritten in the following form:

$$\Delta U_{np.} = (T_2 - T_1)\frac{k}{q} * \ln(\frac{R_1}{R_2}) \qquad (1.12)$$

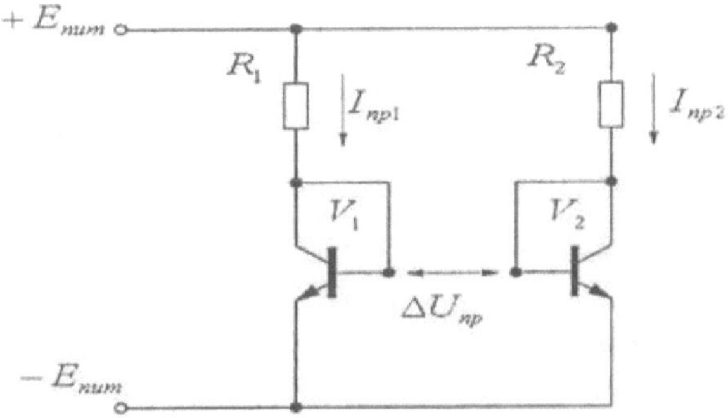

Fig. 1.12. Schematic of a temperature-sensitive element on two identical transistors in diode connection.

Another option to ensure acceptable interchangeability of the temperature-sensitive diode element is to use a multi-emitter transistor structure in the circuit instead of transistor $V1$ [44,45]. Then:

$$\Delta U_{np} = (T_2 - T_1)\frac{k}{q} * \ln(\frac{j_{np2}S_{32}}{j_{np1}S_{31}}) = (T_2 - T_1)\frac{k}{q} * \ln(\frac{j_{np2}}{j_{np1}}) \qquad (1.13)$$

where S_{a1} and $S_{,2}$ are the areas of emitter p-n junctions of transistors V1 and V2, respectively (since the transistors *are the* same in the scheme of Fig. 1.11, then $S_{31}=S_{32})>j_{np1}$ and J_{np2} *are the* current density in the emitter of transistors V1 and V2, respectively.

N- emitter transistor structure and equation (1.14) can be represented:

$$\Delta U_{np} = (T_2 - T_1)\frac{k}{q} * \ln(\frac{S_{32}}{S_{31}}) = (T_2 - T_1)\frac{k}{q} * \ln(n) \qquad 1.14)$$

The circuit diagram is shown in Fig. 1.13.

Since, as shown in the paper, the temperature sensitivity of the primary transducer (Fig. 1.13) is of the order of 0.2 mV/grade at the optimum value of n = 10 B in equation (1.14), it is desirable to manufacture and use it together with the amplifying device V3. Table 1.1 shows the parameters of integrated sensors of STP-35 type.

Table. 1.1

Accuracy at 25°C, DT, °C	Temperature range, °C	Current, mA	Sensitivity, mV/grade	Response time t, sec
STP-35A ±3	-40...+125	0,4...5	10	13

| STP-35B ±2 | -40...+125 | 0,4... 5 | 10 | 13 |
| STP-35C±1 | -40...+125 | 0,4...5 | 10 | 13 |

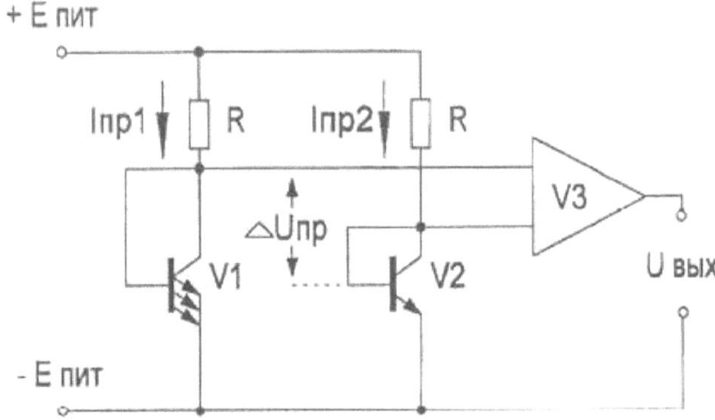

Fig. 1.13. Schematic of a temperature-sensitive element with a multi-emitter transistor structure.

Temperature sensors of the LM3911, LM50, LM60 type are commercially available from National Semiconductor.
The operating principle of the LM3911 temperature sensor is similar to that of a temperature sensor
STP-35 (Fig. 1.14).

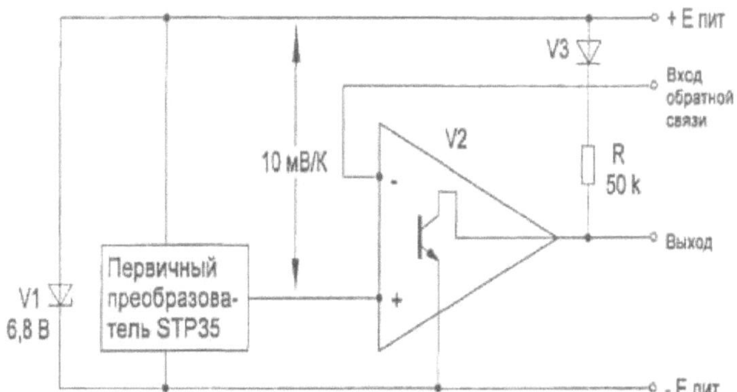

Fig. 1.14. Structure diagram and internal connections of the LM3911 temperature sensor.

When the LM3911 sensor is connected according to the circuit shown in Fig. 1.15, it can be used as a temperature sensor with Celsius scale graduation.

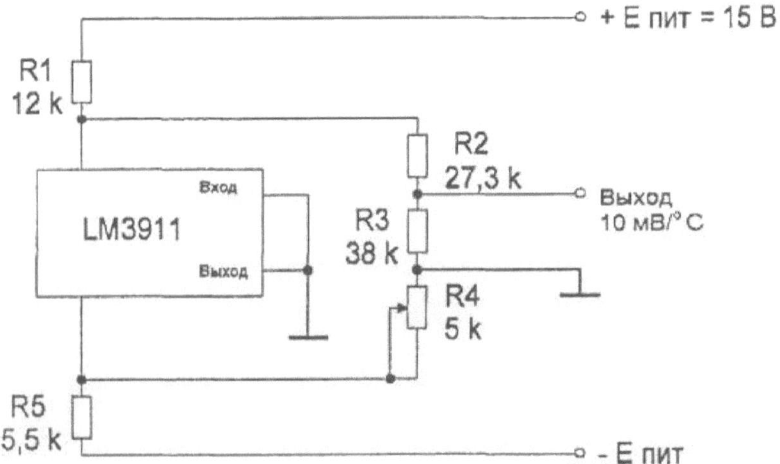

Fig. 1.15. Circuit diagram of the LM3911 sensor with a graduated output of 10 mV/°C,

Fig. 1.16 shows a simplified schematic diagram of the LM60 sensor with internal connections of the primary transducer elements.

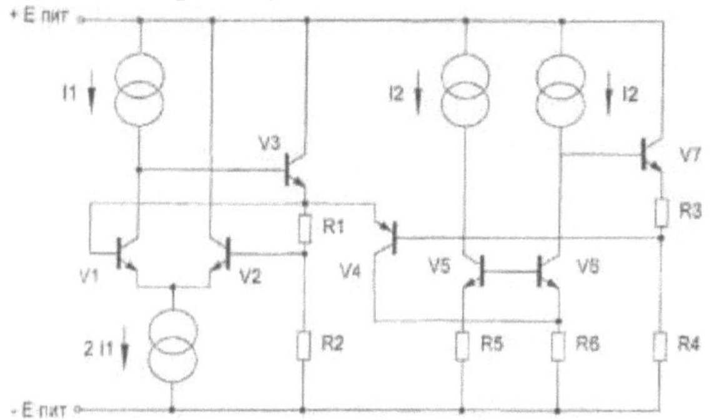

Fig. 1.16. Simplified circuit of LM60 temperature sensor

Fig. 1.17 shows the temperature characteristics of the sensors

LM50 and LM60 temperatures.

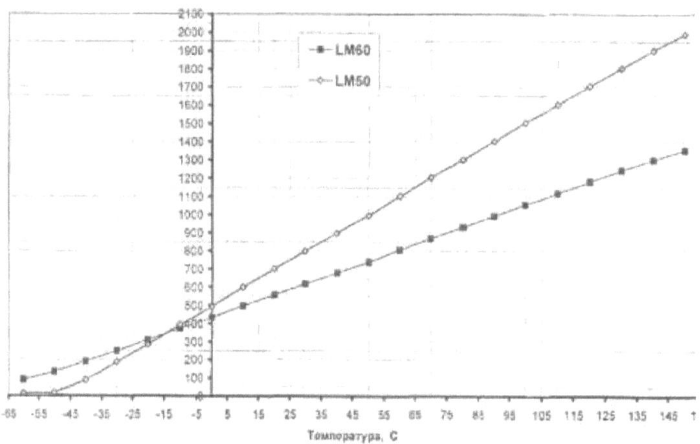

Figure 1.17: Typical dependence of Uвых*∂* LM60 and LM50 on temperature at 10 V.

Dallas Semiconductor (MAXIM) produces integrated temperature sensors based on the primary converter (Fig. 1.13) such as DS18S20, DS18S20, DS18B20, DS1821, DS1822, DS1825 [53, 54], etc. with a digital output, which is provided by an analogue-to-digital converter built into the semiconductor circuit (instead of amplifier V3 shown in Fig. 1.13). Absolute conversion error of DS18S20 sensor in the range of controlled temperatures from -10 °C to +85 °C does not exceed ±0.5 °C and in the range from -55 °C to +100 °C - no more than ±2 °C. The use of digital technology in sensors allows to sharply reduce their power consumption, which provides the possibility of powering such sensors from a device that converts the energy of electrical inductions on the 1-Wire line into the power consumption necessary for the sensors. At the same time, the introduction of digital processing of measuring information in integrated temperature sensors causes serious metrological and operational limitations. For example, the range of measured temperatures (-55°C+125°C) and the range of supply voltages (3.0 V -^-5.5 V) are limited by the capabilities of digital circuits used in the sensor. The results of research show that using the primary transducer shown in Fig. 1.12 and analogue measurement circuitry, it is possible to perform temperature measurements in the range from -55 ° C to +155 ° C with an error of about ±5 ° C with a change in supply voltage from 5 V to 50 V. Fig. 1.17 shows temperature characteristics of commercially available sensors LM50 and LM60 with analogue output, which allow to use supply voltage from 3V to 10V.

1.3.2 Transistor temperature sensor.

Further it is proposed to consider the domestic development (at the level of prototyping) of transistor integrated temperature sensor with analogue output (Fig. 1.18).

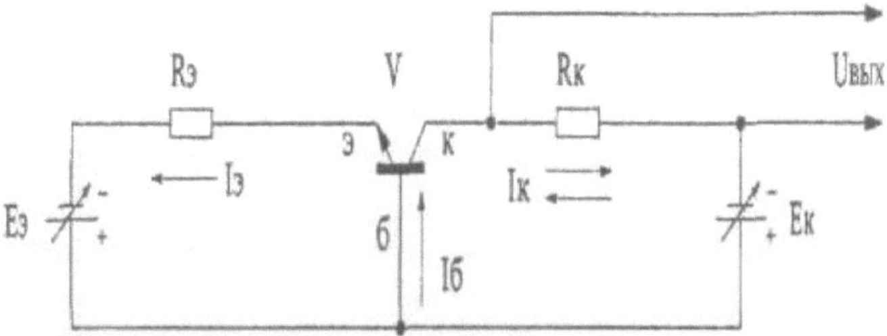

Fig. 1.18. Schematic diagram of transistor inclusion as a converter of temperature into an electrical signal.

The operation of the transistor thermal transducer circuit (Fig. 1.18) is convenient to consider together with the output voltampere characteristics of the transistor, included in the scheme with a common base (Fig. 1.19).

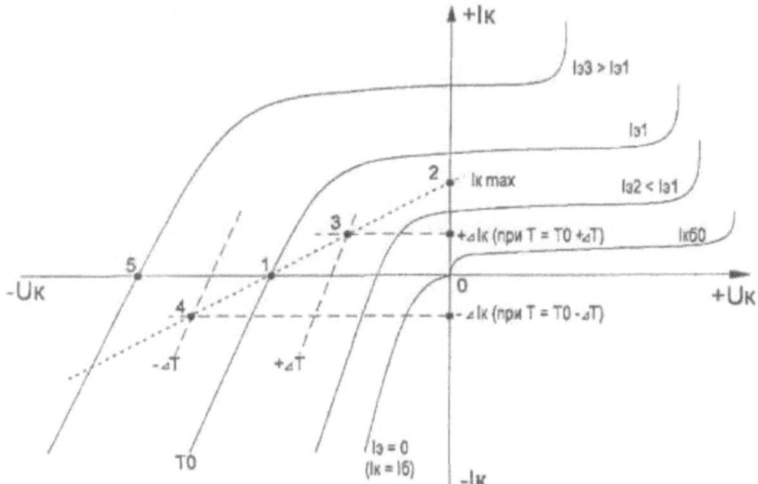

Fig. 1.19. General view of the output voltampere characteristic of a p-p-p transistor switched on in a common-base circuit.

It should be noted that the operating region of this thermal converter takes place only when the transistor is switched on by the scheme with a common base. Fig. 1.19 shows the family of output voltammetric characteristics of the n-p-n transistor, included in the circuit with a common base, at different values of current I_e. At $I_E = 0$ (emitter circuit breakdown), the collector current is the reverse initial collector-base junction current - I_{KBO}, which does not depend on the collector voltage up to the moment of collector-base junction breakdown.

When the polarity of the voltage at the collector-base junction is reversed (in this case from positive to negative), we obtain the voltampere characteristic of the direct biased p-n junction (diode) collector-base. In this case $I_K = I_B$. Setting certain values of the emitter current (I_{e1}, I_{e2}, I_{ez}) we obtain a family of output voltammetric characteristics (Fig. 1.19). In the active region of transistor operation ($U_K > 0$), the voltammetric characteristics are located (Fig. 1.19) parallel to each other and their slope and distance between them does not depend on the value of U_K (up to the breakdown voltage of the collector-base junction), nor on the temperature in the operating temperature range (as follows from the obvious expression $I_K = I_{KBO} + I_{ya}$. since I_E does not depend on temperature, I_{KBO} << I_e, and the transistor force coefficient in a common-base circuit $\alpha \approx 1$). Therefore, this region of transistor operation is not of interest in the design of transistorised thermal converters. Another case is the region in Fig. 1.19, characterised by negative values of U_K, or the region of operation of the transistor with directly biased p-n junctions emitter-base and collector-base. Assuming, in the general case, that the emitter and collector currents are made up of two components (injected and collected), for the family of output voltammetric characteristics of the transistor, included in the circuit with a common base, we can write the expression:

$$I_K = \alpha * I_\mathfrak{z} - I_{KБO} * (e^{\frac{qU_K}{kT}} - 1) \qquad (1.15)$$

Consider in Fig. 1.19 the voltampere characteristic with parameter $I_\mathfrak{z}$, obtained, as well as all output characteristics, at ambient temperature T_o. The point I on the characteristic is determined by the zero value of the collector current, i.e. $I_K = 0$. Then from equation (1.15) we can determine the value of U_K at which $I_K = 0$, namely:

$$U_K = \frac{kT}{q} * \ln(\frac{\alpha I_{\mathfrak{z}1}}{I_{KБO}} + 1) \qquad (1.16)$$

The value of U_K, as follows from equation (1.16), depends on the value of current I_E and can be determined experimentally. If in the circuit (Fig. 1.18) to create the conditions $I_K = 0$ by breaking the collector circuit and $I_e = I_{e1}$, then the circuit provides the ratio $I_{e1} = I_B$, and at the collector terminal creates a potential f_k, equal in magnitude to U_K. Approximately the value of the potential f_k can be determined using a voltmeter with a high impedance input. To create the condition $I_K = 0$, in the scheme of Fig. 1.18, at that $I_e = I_{e1}$ can be another way, not resorting to an interruption in the collector circuit, but by supplying from the collector source E_K voltage on the collector equal in magnitude to the potential f_k, i.e. if $E = f_k$, then $I_K = O$. Experimentally, this condition is easily controlled by measuring the voltage across the resistor R_K, which should be zero, i.e. $U_{BHX} = 0$ and $I_K = I_B$. The results of measuring the source voltage E_K in this case and calculation of U_K using equation (1.16) show that the relation $E_{K=} f_K = U_K$ at $I_K = O$ IS fulfilled with the accuracy necessary for application of this equation in creation of transistor thermal converters. At point 1 (Fig. 1.19), where the condition $I_K = 0$ *and* $I_{el} = I_B$ is fulfilled, the emitter-base p-n junction operates as a diode directly biased and isolated from the collector.

It is known [31] that the diode voltage at constant direct current varies with temperature according to equation (1.16). When the ambient temperature is set $T_{okr} = T_0 - \Delta T$, i.e., when the temperature decreases by ΔT, we obtain a deviation of the characteristic to the left, as shown in Fig. 1.19 by the dotted line to the left of point 1, by the value $\Delta U_K = (TKH) \cdot \Delta T$. A similar shift of the characteristic, but to the right of point 1, as shown in Fig. 1.19 by the dashed line, will occur if the ambient temperature is increased by ΔT, i.e. set $T_{okr} = T_0 + \Delta T$. The load line drawn through point 1 according to the value of resistor R_K intersects the dashed lines discussed above at points 4 and 3. In this case, the value of the output voltage U_{BHX} in the circuit (Fig. 1.18) will have a value according to Fig. 1.19, respectively:

$$U_{B\text{ЫX}} = -\Delta I_K * R_K \tag{1.17}$$

И

$$U_{B\text{ЫX}3} = \Delta I_K * R_K$$

If the condition $\frac{1}{8}I >> \pm\Delta I_K$ is fulfilled, it can be considered that the TKH at the collector is a constant value and that $U_{BHX} = \Delta I_K \cdot R_K \approx (TKH)'\Delta^r \Gamma$, in the range of collector current variation from $-\Delta I_K$ to $+\Delta I_K$. For the initial temperature of the medium T_0 can be set to any value within the operating temperature range, including 0°C. Fig. 1.20 shows the character of change of collector current I_K from the ambient temperature at $T_0 = 0°C$ and from the load resistance R_K for a silicon transistor, switched on according to the scheme of Fig. 1.18.

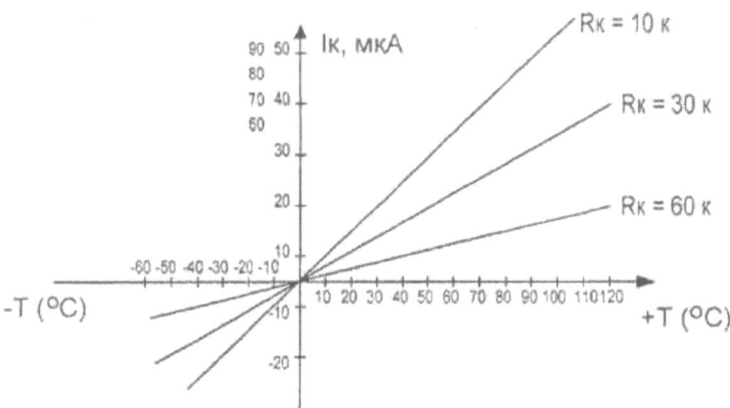

Fig. 1.20. Dependence of collector current 1% on ambient temperature and on load resistance $R_к$ for transistor thermal transducer.

Fig. 1.20. Dependence of collector current $I_к$ on ambient temperature and on load resistance $R_к$ for transistor thermal transducer. Taking into account that the change in the output voltage of integrated temperature sensors, built with the use of silicon resistors and diodes as thermal converters are identical in nature, then, comparing the graphs of the dependence of output voltages on temperature with the graphs in Fig. 1.20 we can make the following conclusions on the operation of the transistor thermal converter:

At a given emitter current, under the condition $I_e >> I_к$ the temperature coefficient of collector current is practically constant in a wider temperature range from -50°C to +150°C in comparison with diode thermal converters and has a high value, about (1 -2) µA/°C, due to small values of collector current, which allows to use the transistor thermal converter without additional amplifying devices.

In the direct current mode it is possible to smoothly adjust the value of collector current within wide limits (including zero value) by changing the voltage Eк, WHICH shifts the collector junction in the forward direction. In this case, the temperature coefficient remains unchanged, the temperature characteristic of the transistor thermocouple (straight line) has positive and negative values of the collector current, corresponding to positive and negative values of the measured temperatures. Such a scale of absolute thermometer is not provided in thermoresistive and diode temperature sensors.

The dependence of the collector current temperature coefficient on the value of the resistor Rк, WHICH does not depend on temperature, makes it possible to adjust the sensitivity of the transistor thermal transducer regardless of the material and technology of its manufacture, unlike its semiconductor analogues. Thus, this paper considers the most promising semiconductor thermal converters and methods of creating semiconductor integrated temperature sensors on their basis. However, it should be noted that, despite the significant contribution of the domestic industry to the development of semiconductor thermal converters, integrated temperature sensors are serially mastered and mastered only by foreign firms.

1.4 Methods of temperature measurement with diode sensors.

Method 1. The considered relationship between U and $/$ allows us to directly determine the absolute temperature T by measuring the voltampere characteristic of a semiconductor diode. From relations (1.5), (1.6) it follows:

$$I - I_{нас} = I_o * \exp(-\frac{\varepsilon_g}{kT} + \frac{eU}{kT})$$
(1.18)

Usually in p-n-junctions the forward current can be many orders of magnitude larger than the saturation current $/_{nas}$ (for example, in low-power silicon diodes $/_{nas} \sim 10\text{-}9\,A,\ I \sim K)\text{-}\ A$). In such a case in the left part of the relation (1.5) we can neglect $/_{nas}$. Then $ln/$ will be a linear function of voltage U:

$$\ln I = \ln I_o + (-\frac{\varepsilon_g}{kT} + \frac{eU}{kT})$$
(1.19)

Having plotted the experimental graph of the dependence $ln/ = f(U)$ for direct current, we can find the angular slope coefficient of the graph

$$M = \frac{d \ln I}{dU} = \frac{e}{kT}$$
(1.20)

from which the temperature T is determined:

$$T = \frac{e}{kM}$$
(1.21)

This method is one of the few that allows measuring absolute temperature without preliminary thermometer graduation, but it is not practically applied due to its labour-intensive nature.

Method 2. Temperature measurements can be realised by using a p-n junction in the direct current mode in direct switching. From the formula (1.5) in the same approximation $I >> I_{nas}$ it follows that at $I = const$ the diode voltage U is a linear function of temperature T and can be a convenient thermometric parameter:

$$U = \frac{kT}{e} * \ln \frac{I}{I_o} + \frac{\varepsilon_g}{e} = AT + B$$
(1.22)

where constants A and B do not depend on temperature.

The constant A *is* negative because the resistance of the semiconductor decreases with increasing temperature and, therefore, at constant current the voltage across the sample also decreases. Since the value of I_o and ε_g are not known in advance, the coefficients A and B must be determined empirically when calibrating the sensor and can then be used to measure the temperature. The constant B also allows one to find ε_g, the forbidden band width of the semiconductor. Semiconductor thermometers usually use silicon or germanium diodes as the temperature sensor.

CHAPTER 2

MANUFACTURING TECHNOLOGIES FOR THREE-ELECTRODE TEMPERATURE SENSORS

2.1 Technological parameters of manufacturing of three-electrode temperature sensors

Currently, electronic thermometers use thermodiodes, thermotransistors, resistor and integrated thermosensors for temperature measurement. Among other things, Uzbekistan is working on the development of temperature sensors in the form of silicon thermoresistors with compensated base area: using diffusion technology by doping with transition metal impurities (manganese, nickel, etc.), and by thermal radiation doping of silicon. In addition, based on integrated circuits on complementary transistors consisting of several dozens of transistors, a number of thermal sensors capable of monitoring the temperature of various objects have been proposed [21].

A common disadvantage of known semiconductor sensors is their low accuracy of temperature measurement. The accuracy of temperature measurement in semiconductor sensors is determined by a combination of many factors, most of which is a consequence of the strong dependence of temperature sensitivity of these structures on technological and material parameters, which due to non-ideal reproducibility of technological production of semiconductor devices have a fairly large variation. In addition, the presence of temperature dependence of such parameters as non-ideality coefficient or series resistance of the structure leads to a significant decrease in measurement accuracy.

To create a temperature sensor devoid of these disadvantages, we propose a new structure and design of the temperature sensor, which is a three-electrode silicon structure with a depleted base region.

We can see a study of the thermal sensitivity of silicon structures with depleted base area, which uses the full base area depletion voltage (U_o) as a measurement parameter.

The type of analytical relations linking structural and static parameters is determined by the nature of impurity distribution in the base region.

It is known from the literature [2,3,21] that in structures similar to the structure of the proposed thermosensor, i.e., the structure of p+-p-p-type, the distribution of impurity concentration in the base region is parabolic.

The parabola is symmetric with the maximum located in the middle of the base. However, in these works, the considered p+-p-p-type structures had a rather large base thickness, $5 \div 10$ μm, while in the proposed thermal sensor the base thickness is approximately an order of magnitude smaller.

We have attempted to find computationally the impurity concentration distributions in the base of the proposed temperature sensor.

For this purpose, the technological process for the formation of the proposed temperature sensor was dissected step by step.

A). Study of impurity concentration distribution in p-n type epitaxial-planar structure under specific modes of structure fabrication.

Diffusion during the growth of autoepitaxial layers at the substrate-epitaxial layer boundary was calculated using Eq:

$$N(x,t) = \frac{N_{10}}{2}\left(1 - erf\frac{x}{2\sqrt{Dt}}\right) + \frac{N_{20}}{2}\left(1 + erf\frac{x}{2\sqrt{D_2 t}}\right), \qquad (2.1)$$

where $N(x, t)$ is the finite impurity distribution.

N_{10} - initial concentration of impurity in the layer, in case of homogeneous.

N_{20} - initial concentration of impurity in the substrate.

For cases: a) where a low resistivity epitaxial film is grown on a high resistivity substrate;

b) when a high resistivity epitaxial film is grown on a low resistivity substrate.

Calculations of this type and the construction of impurity concentration are available in the literature. Of interest is the case when the resistivity of the substrate and the layer are of the same order. Such calculations are not found in the literature, and the proposed temperature sensor has exactly such a structure of epitaxial layer and substrate.

We have made calculations for $R_{PODL} \leq R_{EP} \leq 2 P_{POD}$ where R_{EP} is the resistivity of the epitaxial film, p_{PODL} is the resistivity of the substrate.

Specific modes of fabrication of p-n-type epitaxial-planar structure were used in the calculation. An n-type epitaxial layer with resistivity $r_{EP} = /0.6 \div 0.9/ohm.cm$ was grown on a p-type silicon substrate with $R_{PODL} = 0.5$ ohm^cm. at $T = 1250^0$ C; $t = 5$ min.

Б). Change of impurity distribution near the epitaxial layer-substrate interface after heat treatment.

The technological process of manufacturing the proposed thermal sensor consists of a number of operations in which the structure "epitaxial layer-substrate" is subjected to heat treatment. These operations are as follows:

a) oxidation

б) the first diffusion of boron-separation. Diffusion is two-stage, i.e. first boron driving and then distillation.

The oxidation process is also carried out during distillation.

в) The 2nd boron diffusion is carried out to obtain the upper limiting p+ layer. Diffusion is also two-stage /diffusion, distillation/.

The distillation mode is chosen differently depending on the thickness of the epitaxial layer

(c) Phosphorus diffusion to obtain n+ at contact areas. Diffusion is two-stage /diffusion, distillation/.

In order to obtain the final distribution in the channel after all technological operations it is necessary to take into account all diffusion and oxidation processes.

B). Impurity concentration distribution in the epitaxial planar structure of the p-n junction. The location of the "top" p-n junction formed in the epitaxial layer by the second boron diffusion is determined by the mode of this diffusion. In our case, a two-step diffusion process is used.

Initially, a short-term diffusion is carried out, described by Eq:

$$N(x,t) = N_o\left(1 - erf\frac{x}{2\sqrt{Dt}}\right), \qquad (2.4)$$

where N_o is the surface concentration of the impurity.

The thin layer obtained in the first stage serves as a limited source for the second stage. If on the surface there is a limited source with impurity concentration per unit surface S, the solution has the form:

$$N(x,t) = \frac{S}{\sqrt{\pi_1 Dt}}e^{-\frac{x^2}{4Dt}}, \qquad (2.5)$$

However, the layer obtained in the first diffusion stage can be considered as a limited source for the second stage, provided that the magnitude of the product Dt for the second diffusion /diffusion/ stage is significantly larger than for the first diffusion /diffusion/ stage.

In this case, the value S is the total number of impure effluents per $1cm^2$ and introduced in the 1st stage /pens/.

$$S = \int_0^\infty N(x)dx = \int_0^\infty N_o\left(1 - erf\frac{x}{2\sqrt{Dt}}\right)dx = \frac{2N_o}{\sqrt{x}}\sqrt{Dt} \qquad (2.6)$$

After substitution we obtain:

$$N(x,t) = \frac{2N_o}{\pi}\sqrt{\frac{Dt}{D't'}}e^{-\frac{x^2}{4Dt}} \qquad (2.7)$$

where $D't'$ - refers to the second stage of diffusion /diffusion/.

The difference D and D' is caused by carrying out these two stages of diffusion at different temperatures. If the duration of the second stage is short in comparison with the first stage, the assumptions that the diffusion layer formed as a result of enclosure can be considered as a limited source are incorrect.

The solution in this case is of the form:

$$N(x,t,t') = N_o \frac{2}{\sqrt{\pi}} \int_{\beta}^{\infty} e^{-m^2} erf(\alpha m)dm \qquad (2.8)$$

Where

$$\alpha = \sqrt{\frac{Dt}{D't'}} \qquad \beta = \frac{x^2}{4(Dt + D't')} \qquad (2.9)$$

In the literature, there is a solution of such *an* integral for different values of *a* and ß.

In our case, the dispersal duration is not much longer than the corral duration t' $\geq 2t$.

Namely (a) t5 = 50 min; $Dt = 2.1*10^{\wedge 1} \circ cm^2$; $D't' = 12*10^{\wedge 1} \circ cm^2$

b) t5=14°min; $Dt = 2.1*10^{\wedge 1} \circ cm^2$; $D't' = 33.6*1^{\circ\wedge 12} cm^2$

using the existing solution, we obtain the distribution of diffusion boron in the epitaxial layer. $N(x)$

Table 2.1 summarises these data for two different surface concentrations - N_o /The *x* surface of the epitaxial layer is taken as the starting point/.

Using the data of Tables 2.1 and 2.2, the concentration distribution of impurities in the epitaxial planar structure p^+-n-p can be obtained. Fig.2.12.2 shows such distributions for different epitaxial layers and under different diffusion modes.

$N_2 = n$ are the resulting concentrations in Fig.

N1- concentration of boron impurity in the epitaxial layer pro diffused as a result of the second diffusion.

N is the total concentration of impurities in the p-n-p structure.

Distribution of boron impurity concentration in the epitaxial layer (after 2nd boron diffusion).

Table 2.1.

N_o cm^{-3}	t5 MUH	ß	a	x micron	AT. -3 N cm^3
				0,5	$5,6*10^{18}$
				1,0	$1,4*10^{18}$
				1,5	$1,28*10^{17}$
$2*10^{19}$	50	0.564*10 8	0,41	1,68	$4,6*10^{16}$
				1,99	$5,4*10^{19}$
				2,25	$6,98*10^{14}$
				2,49	$7,44*10^{13}$
				0,51	$1,4*10^{19}$
5,1019	50	0.564*10 8	0,41	1,0	$3,6*10^{18}$
				1,5	$3,2*10^{17}$
				1,68	$1,15*10^{17}$

				1,99	$1,35*10^{16}$
				2,25	$1,64*10^{15}$
				2,49	$1,86*10^{14}$
$2,10^{19}$	140	$1.43*10\ 8$	0,25	0,55	$4,52*10^{18}$
				1,0	$2,8*10^{18}$
				1,5	$9,6*10^{17}$
				2,07	$2,54*10^{17}$
				2,38	$9,2*10^{16}$
				2,69	$3,4*10^{16}$
				3,16	$3,84*10^{15}$
$5*10^{19}$	140	$\overset{2x}{1.43*10\ 8}$	0,25	0,5	$1,13*10^{19}$
				1,0	$7*10^{18}$
				1,5	$2,4*10^{18}$
				2,07	$6,35*10^{17}$
				2,38	$2,3*10^{17}$
				2,69	$8,5*10^{16}$
				3,16	$9,6*10^{15}$

Analysing Figures 2.1. -2.2 we can draw the following conclusions:

1. The distribution of impurity concentration in the regions /"upper p-layer"/ and /base/, /"lower p-layer"/ is inhomogeneous.

2. In the base region, the distribution of the resulting impurity has the form of an asymmetric parabola with a maximum located closer to the "upper" transition.

3. The position of the maximum of the parabola changes slightly in the depending on the resistivity of the epitaxial layer and the mode of the second diffusion /see Table 2.2/.

Dependence of the position of the parabola maximum on the resistivity of the epitaxial layer and the mode of the 2nd boron diffusion.

Table 2.2.

$B_{E,C,}$ /om.cm./	N_o $1/cm^3$	$t5$ /min	x $mfil$
0,6	$2* i9_w$	50	$0,38\ W_{jt}$
0,6	$5* i9_w$	50	$0,49\ W_{jt}$
0,6	$2*10^{19}$	140	$0,45\ W_{jt}$
0,7	$2*10^{19}$	50	$0,3\ w_{if}$
0,7	$5*10^{19}$	50	$0,39\ W_{if}$
0,7	$2*10^{19}$	140	$0,4\ W_{if}$
0,8	$2*10^{19}$	50	$0,25\ W_{if}$

0,8	$2*10^{19}$	140	$0,29\ W_{jt}$
0,9	$2*10^{19}$	50	$0,23\ W_{if}$
0,9	$2*10^{19}$	140	$0,31\ w_{jt}$

In Table 2.2: $_{No}$ - surface concentration of boron impurity in the 2nd

diffusion.

t - boron dispersion time /2 boron diffusion/.

[a]τn is the position of the maximum of the parabola.

W_{ii} - width of the parabola /base/.

Thus, our calculation shows that for the specific p^{+}-n-p structure considered by us, which corresponds to the proposed thermal sensor, the distribution of impurity in the n-region can be considered parabolic, but the parabola is asymmetric with the maximum shifted to the "upper" p-layer.

For the study, 2 types of silicon p-n structures (temperature sensor crystals) were fabricated, consisting of an n-type phosphorus-doped epitaxial layer with a carrier concentration of $5\text{-}10^{15}$ cm^{-3} at a thickness of 3.2±0.2 μm in type I samples and 2.3±0.2 μm in type II samples, which is grown on a p-type silicon substrate oriented in the (100) plane and doped with boron, with a carrier concentration of $3\text{-}10^{16}$ cm^{-3} and a thickness of 230±20 μm. On a part of the surface of the n-type epitaxial layer, an additional heavily doped p+-type region with a carrier concentration of $3\text{-}10^{18}$ cm^{-3} with a thickness of 2.0±0.2 μm in the samples of type I and 1.2±0.2 μm in the samples of type II was formed by boron diffusion, which is technologically short-circuited with the p-type substrate through diffusion regions of p+-type, which are formed along the edge of the structure. The length of the p+-type region was 2.5 μm in type I samples and 2.0 μm in type II samples. In addition, two strongly doped n-type regions with thicknesses of 2.5±0.2 μm in type I samples and 1.8±0.2 μm in type II samples were formed by phosphorus diffusion. Contacts to these regions were formed by sputtering an Al layer, 0.5±0.2 μm thick. The distance between these n-rnim regions was 20 μm in type I samples and 14 μm in type II samples.

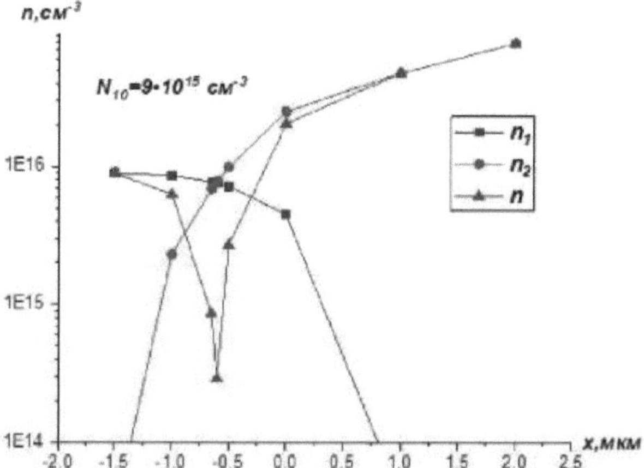

Fig.2.1 Impurity distribution profile in the epitaxial layer on the substrate.

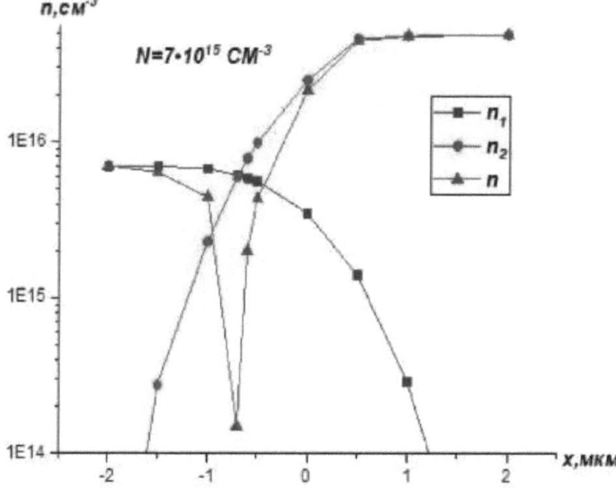

Fig.2.2 Impurity distribution profile in the epitaxial layer on the substrate.

As a result, two rectifying junctions, one with an n-p$^+$ -junction and the other with a p-n junction, are formed in the finished structure shown in Fig. 2.3, which are connected in parallel, since the p$^+$ -type region is short-circuited with the p-type region. With a total crystal size of 0.46x0.46·10^{-2} cm^2 in type I samples and 0.5x0.5·10$^{^2}$ cm2 in type II samples, the active area of the p-n junction in both samples was - 2.36- 10^{-4} cm2 and the p-p$^+$ -junction also in both samples was - 0.2·10$^{^4}$ cm2.

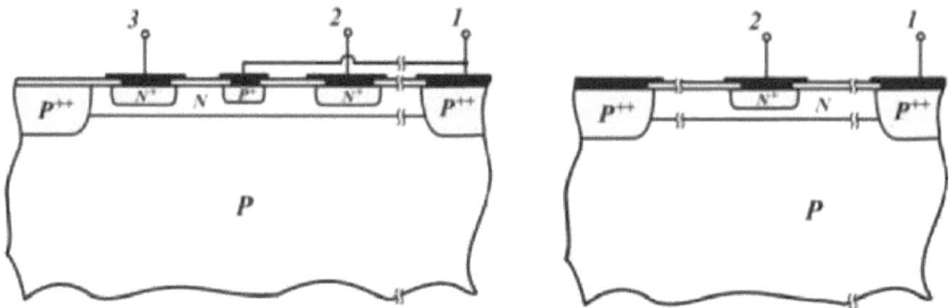

Fig. 2.3. Structural diagram of the investigated three-electrode silicon structure. To facilitate the study, these structures were mounted on metal-glass enclosures of the KT-1-12 type (30 pcs. of type I and 30 pcs. of type II samples), in addition, the hull-less versions were also produced (20 pcs. samples of type I), Fig.2.4.

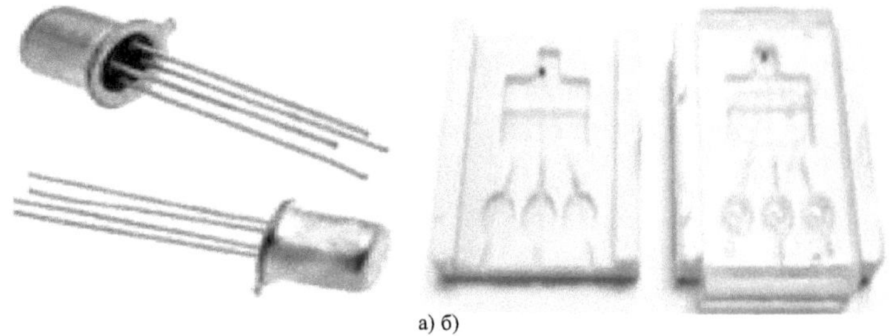

Fig. 2.4. Samples in the housing type a KT-1-12 (a) and without housing (b)

2.2 Volt-farad characteristic of thermal sensors and relationship with structural parameters.

Considering the relationship between structural and static parameters of thermal sensors, we note the essential role of volt-farad characteristics. In this paragraph, we will consider the methodology for determining the thickness and length of the device base using the volt-farad characteristics $C_{3n} = f(U_3)$ and $C_{3c} = f(U_3)$, i.e., from the input and output characteristics. The specific capacitance of a p-junction can be represented as the capacitance of a concentrator in which the distance between the plates is equal to the thickness of the space charge layer h, and the dielectric constant of the space between the plates is ε, i.e.:

$$C = \frac{\varepsilon}{h} \qquad (2.9)$$

The specific capacitance of the p-p junction in this representation clearly does not depend on the nature of the charge distribution in the base - $p(y)$ and on the voltage drop in the space charge layer, but since the value of h is a function of $p(h)$ and U, the specific capacitance-function of the coordinate/ since the thickness of the space charge layer varies in the direction from contact 1 to contact 2/.

The capacitance of the p-junction and the course of the volt-farad characteristic is determined both by the law of distribution of impurities in the channel and by the external voltage applied between the contacts. The capacitances between pins 1-3 of c_{13} and pins 2-3 of c_{23} depend on the voltage across pin 3 at the initial (Fig.2.5) fixed voltage value / $U_c < U_0$ /. At voltages below the depletion voltage these capacitances are almost equal to each other.

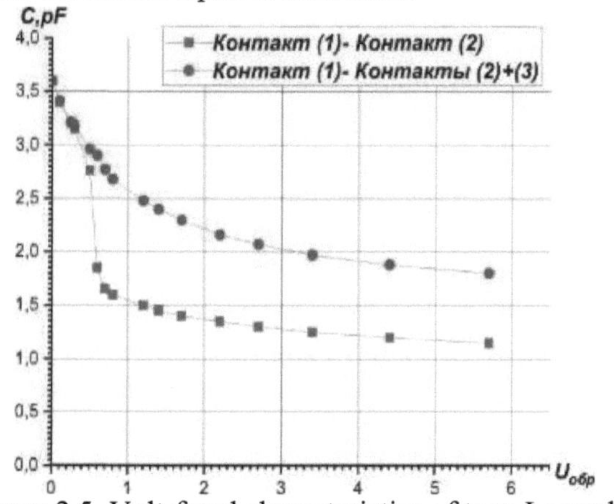

Figure 2.5. Volt-farad characteristics of type I samples

When depletion is reached, there is a sharp change in the capacitances and at voltages higher than the depletion voltage, c_{13} and c_{23} are markedly different from each other. This pattern of change in the values of c_{13} and c_{23} can be explained as follows: outside the depletion region, both capacitances are determined by the full area of the p-junction, when the depletion voltage is reached, the space charge region covers the base region and cuts off from pin 2. Under these conditions, the capacitance of c_{13} and c_{23} are determined only by the part of the area bordering directly on contact regions 1 and 2, and since these parts are not the same, the capacitances of c_{13} and c_{23} are different. Therefore, the value of the "Break" on the volt-farad characteristic (Fig. 2.6), corresponding to the depletion voltage of the base region, can be used to determine the thickness of the base.

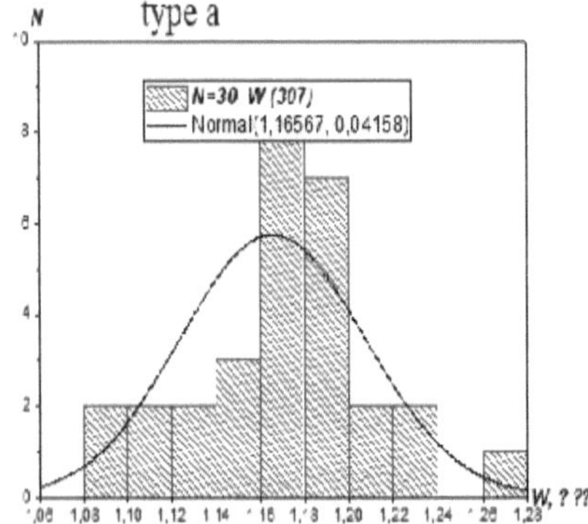

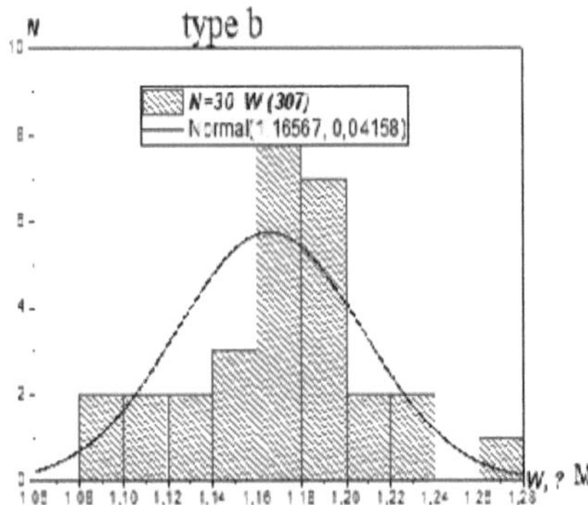

Fig. 2.6. Volt-farad characteristics in different connection modes (a), histogram of basic density distribution of silicon temperature sensors type I (b) and type II (c).

There are a number of works in the literature, which derive analytical expressions relating the p-1 junction capacitance to the p-region potential. However, these works are done for a uniform distribution of impurities in the base. We have attempted to derive an analytical expression relating the p-1-junction capacitance to the p-layer potential for the case of parabolic distribution of the resulting impurity in the channel /which seems to be the most probable approximation for the proposed thermal sensors/. At the same time, the p-layer transition was considered to be: a) sharp, b) smooth.

The expressions are of the following form:

$$A)\ C_{(\upsilon)} = 16\varepsilon \frac{H_k L_k}{W_k} \frac{\left(1+\upsilon^{\frac{1}{2}}\right)^3}{\left(1+2\upsilon^{\frac{1}{2}}\right)^2 \left(1-2\upsilon^{\frac{1}{2}}\right)^2}$$

$$Б)\ C_{(\upsilon)} = \frac{15}{7}\varepsilon \frac{H_k L_k}{W_k} \frac{\left(3-7\upsilon^{\frac{2}{3}}+7\upsilon^{\frac{5}{3}}-3\upsilon^{\frac{7}{3}}\right)\left(1-\upsilon^{\frac{2}{3}}\right)}{\left(2-5\upsilon+3\upsilon^{2}\right)^{\frac{5}{2}}}$$

Where $\upsilon = \dfrac{U_3}{U_o}$ U_3 is the p-layer voltage , U_o is the depletion voltage.

H_k - base width; L_k - base length; W_k - base thickness. At the "break" point at

$$\upsilon = \frac{U_3}{U_o} = 1$$

volt-farad characteristic expressions are eliminated:

34

A) $C = \dfrac{16*8}{5*9} \varepsilon \dfrac{H_k L_k}{W_k}$

Б) $C = \dfrac{25}{7} \varepsilon \dfrac{H_k L_k}{W_k}$

Specific capacitance of p-p junction

A) $\dfrac{C}{H_k L_k} = \dfrac{16*8}{5*9} \dfrac{\varepsilon}{W_k}$

Б) $\dfrac{C}{H_k L_k} = \dfrac{25}{7} \dfrac{\varepsilon}{W_k}$

Thus, expressions a and b can be used to determine the
base thickness W_k.
This method can be used to make a comparative assessment on the thickness of
the base of the same type of temperature sensor.
Individual transistors of this type can also be different and the base length L_k
/base width H_k usually holds up reasonably well and
can be considered constant/. To calculate the base length, additional
measurements are required, Figs. 2.7-2.8.

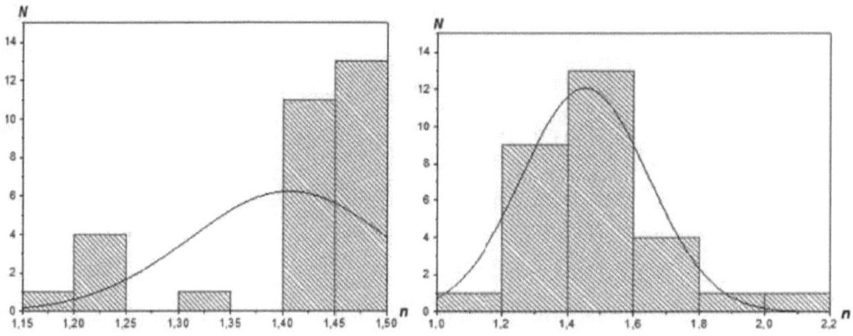

Figure 2.7. Histogram of the distribution of the ideality coefficient
silicon temperature sensors type I (a) and type II (b)

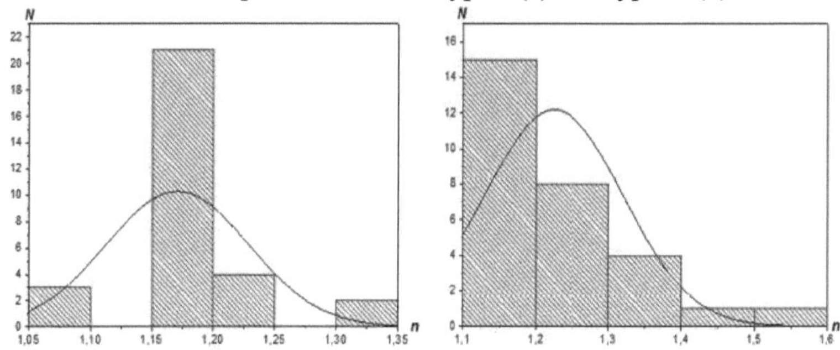

Figure 2.8.

Histogram of the distribution of the ideality coefficient of silicon temperature sensors type I (a) and type II (b).

2.3 Evaluate opportunities to improve the accuracy of the temperature sensor.

The present work was also devoted to determining the dependence of the accuracy of temperature measurement by semiconductor sensors on the small technological variation of the values of the non-ideality coefficient, as well as the bias current stabilisation coefficient.

As is well known, the expression for the thermal sensitivity of conventional diode temperature sensors can be derived by using, directly, the following diode equation:

$$I = A * J_S * \left(\exp\left(\frac{qV}{nkT} \right) - 1 \right) \tag{2.10}$$

where A is the area of the structure, $_{JS}$ is the saturation current, n is the non-ideality coefficient, k is the Boltzmann coefficient, T is the temperature, q is the electron charge. However, in contrast to the traditional approach [5], we have transformed equation (2.10) into the following equation when deriving the expression for thermosensitivity by introducing the notation

$$I_S^* = A * J_S * \exp\left(\frac{qV_k}{nkT} \right) \qquad V_k \gg \frac{nkT}{q},$$

and taking into account:

$$I = I_S^* * \exp\left(-\frac{q(V_k - V)}{nkT} \right), \tag{2.11}$$

One advantage of this conversion is the independence of JS on temperature, unlike JS which has an exponential dependence on inverse temperature.

From equation (2.11) we can directly find the voltage falling across the diode contacts:

$$V = V_K - \frac{nkT}{q} \ln\left(\frac{I_S^*}{I} \right), \tag{2.12}$$

or including

$$V_K = \frac{E_G}{q} - \frac{kT}{q} \ln\left(\frac{N_C N_V}{N_D N_A} \right),$$

will get:

$$V = \frac{E_G}{q} - \frac{kT}{q} \ln\left(\frac{N_C N_V}{N_D N_A} \right) - \frac{nkT}{q} \ln\left(\frac{I_S^*}{I} \right), \tag{2.13}$$

As can be seen from equation (2.13), the voltage falling on the diode contacts is the sum of several terms, and depends not only on the material parameters of the structure, but also has a strong dependence on the mechanism of current transfer and its changes with temperature change, which in addition are subject to the influence of external factors that lead to a large scatter of measurement values. In particular, if the bias current of the temperature sensor has a spread equal to ΔI, then the voltage falling on the diode contacts can be written in the following form:

$$V = \frac{E_G}{q} - \frac{kT}{q}\ln\left(\frac{N_C N_V}{N_D N_A}\right) - \frac{nkT}{q}\ln\left(\frac{I_s^*}{I+\Delta I}\right) = \frac{E_G}{q} - \frac{kT}{q}\ln\left(\frac{N_C N_V}{N_D N_A}\right) - \frac{nkT}{q}\ln\left(\frac{I_s^*}{I}\right) -$$
$$- \frac{nkT}{q}\ln\left(1+\frac{\Delta I}{I}\right) = V\big|_{\Delta I=0} - \frac{nkT}{q}\ln\left(1+\frac{\Delta I}{I}\right) \cong V\big|_{\Delta I=0} - \frac{nkT}{q}\frac{\Delta I}{I}$$

(2.14)

In the same way it is possible to deduce the dependence of the voltage falling on the contacts of the diode on a small technological variation of the non-ideality coefficient values:

$$V = \frac{E_G}{q} - \frac{kT}{q}\ln\left(\frac{N_C N_V}{N_D N_A}\right) - \frac{(n+\Delta m)kT}{q}\ln\left(\frac{I_s^*}{I}\right) = \frac{E_G}{q} - \frac{kT}{q}\ln\left(\frac{N_C N_V}{N_D N_A}\right) - \frac{nkT}{q}\ln\left(\frac{I_s^*}{I}\right) -$$
$$- \frac{\Delta nkT}{q}\ln\left(\frac{I_s^*}{I}\right) = V\big|_{\Delta I=0} - \frac{\Delta nkT}{q}\ln\left(\frac{I_s^*}{I}\right)$$

(2.15)

As can be seen from Figs. 2.9-2.10, the presence of technological variation of such parameters as the non-ideality factor and the bias current stabilisation factor leads to a significant decrease in measurement accuracy. In particular, at room temperature and a bias current of 1 µA, a 1% variation of the bias current leads to a measurement error of ±0.11° C, while a 10% variation leads to a measurement error of ±1.08° C.

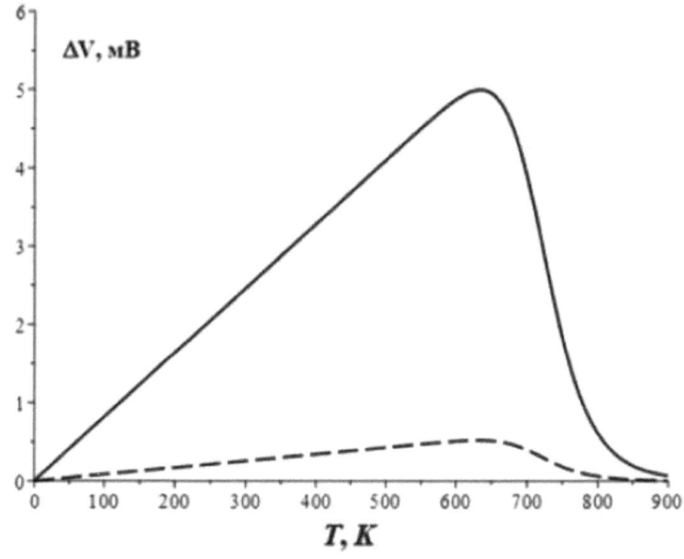

Fig. 2.9. Dependence of the variation of the voltage incident on the diode contacts at different bias current variations: solid line -10%, dashed line - 1%.

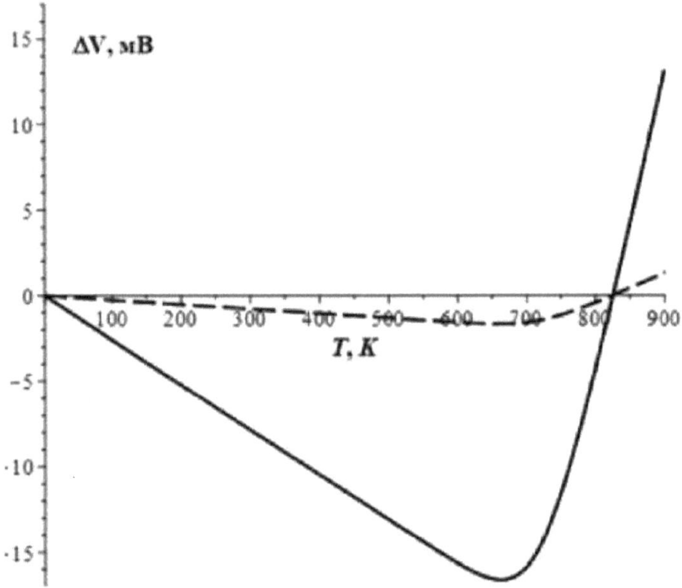

Fig. 2.10. Dependence of the variation of the voltage incident on the diode contacts at different scatterings of the non-ideality coefficient: solid line - 10%, dashed line - 1%.

CHAPTER 3

. ELECTROPHYSICAL PROPERTIES OF THE PROPOSED TEMPERATURE SENSOR

3.1. Investigation of electrophysical properties of laboratory samples of temperature sensors

Concentration of charge carriers and the character of its distribution along the thickness were measured by measuring the volt-farad characteristics of the structures using a capacitance meter L2-28. The measurement error was not more than 0.5%. The depth of the n-p+-junction was controlled on the test structures by measuring the clamping voltage of the n-p+- and p-n-junctions and, if necessary, additional boron dispersion was carried out to bring the clamping voltage of the junctions to the required value.

Temperature studies were carried out using a special cryostat in which liquid nitrogen was used as a refrigerant, and the temperature regime was set using a heater mounted in the sample holder placed inside the measuring chamber. The set temperature in the range from -180° C to 180° C was measured by a chromel-alumel thermocouple and maintained by a thermoregulator 2TRM1. The temperature maintenance error did not exceed 0.1° C. The investigations showed (Fig.3.1) that the p-n junctions in the samples are sharp, and also the samples have a very small variation in the doping degree of the base region.

The voltammetric characteristics of the laboratory samples were also investigated and are shown in Fig. 3.1.

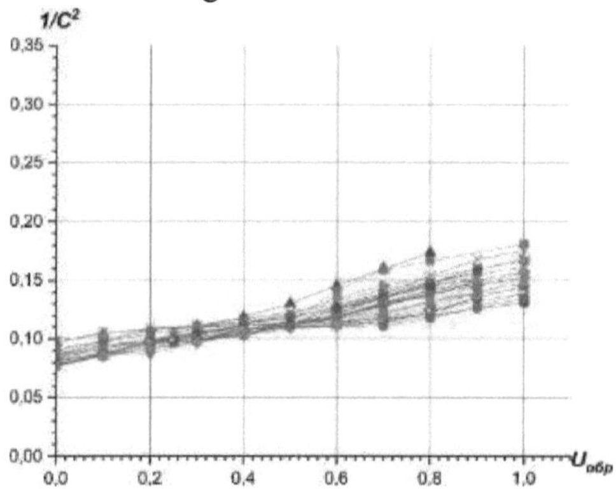

Fig. 3.1a. Volt-farad characteristics of type I samples.

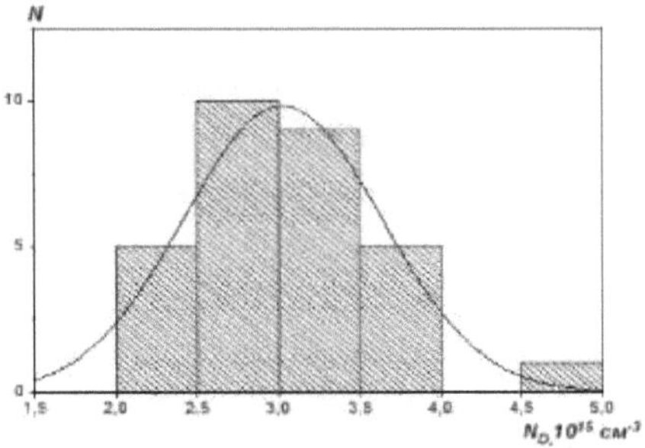

Fig. 3.1a. Histogram of impurity concentration distribution in the base of silicon temperature sensors of type I.

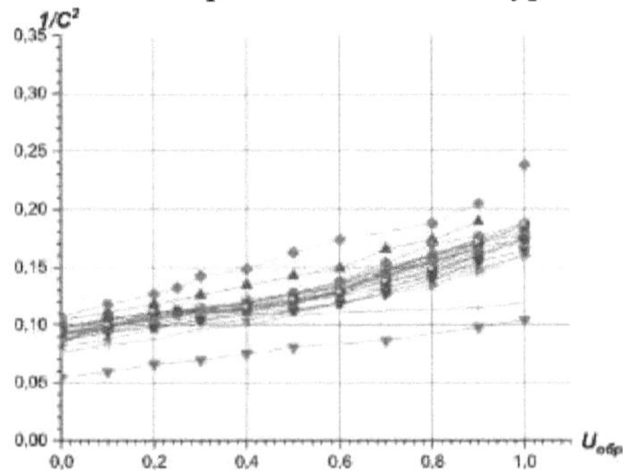

Fig. 3.1b. Volt-farad characteristics of type II samples.

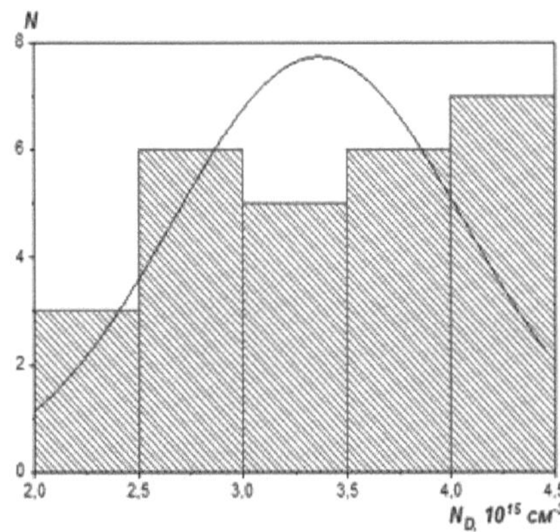

Fig. 3.1b. Histogram of impurity concentration distribution in the base of silicon temperature sensors of type II.

3.2. Comparative analysis of the proposed temperature sensors with traditional analogues

A comparative analysis of the thermal sensitivity of our proposed thermosensor

40

with the sensitivity of conventional diode thermosensors was carried out.
As is well known, the expression for the thermal sensitivity of conventional diode temperature sensors can be derived by using, directly, the following diode equation:

$$I = A \cdot J_s \cdot \left(\exp\left(\frac{q(V - IR_S)}{nkT} \right) - 1 \right),$$
(3.1)

where A is the area of the structure, J_S is the saturation current, RS is the series resistance, n is the non-ideality factor, k is the Boltzmann factor, T - temperature, q is the electron charge.

However, in contrast to the traditional approach, we have transformed equation (3.1) into the following equation when deriving the expression for the

thermosensitivity by introducing the notation $\quad I_S^* = A \cdot J_s \cdot \exp\left(\frac{qV_K}{nkT} \right)$

considering : $\quad V_K \gg \dfrac{nkT}{q}$

$$I = I_S^* \cdot \exp\left(-\frac{q(V_K - (V - IR_S))}{nkT} \right),$$
(3.2)

where V_K *is the* contact potential difference.

One of the advantages of this transformation is the independence of J^*_S from temperature, in contrast to $J_,$, which has an exponential dependence on the inverse temperature. In addition, this transformation, as will be shown below, allows us to derive the expression of thermal sensitivity for a traditional temperature sensor in a form similar to the expression obtained for our proposed temperature sensor, which greatly simplifies their comparative analysis.

From equation (3.2) we can directly find the voltage incident on the

contacts of the diode: where E_G is the width of the forbidden zone; N_c , N_V are the effective density of states in the conduction band and valence band, respectively; N_D , N_A are the concentrations of donors and acceptors, respectively.

$$V = V_K - \frac{nkT}{q} \ln\left(\frac{I_S^*}{I} \right) + I \cdot R_S,$$
(3.3)

or including $\quad V_K = \dfrac{E_G}{q} - \dfrac{kT}{q} \ln\left(\dfrac{N_C N_V}{N_D N_A} \right),$ will get:

$$V = \frac{E_G}{q} + I \cdot R_s - \frac{kT}{q} \ln\left(\frac{N_C N_V}{N_D N_A}\right) - \frac{nkT}{q} \ln\left(\frac{I_s^*}{I}\right), \qquad (3.4)$$

As can be seen from equation (3.4), the voltage falling on the diode contacts is the sum of several terms, and depends not only on the material parameters of the structure, but also has a strong dependence on the mechanism of current transfer and its changes with temperature change, which in addition are subject to the influence of external factors that lead to a large scatter of measurement values (Fig. 3.2a).

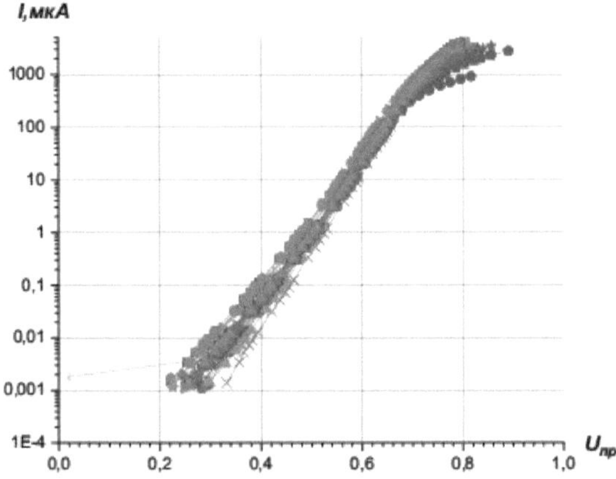

Fig. 3.2a. Volt-ampere characteristics of type I samples.
In the proposed temperature sensor with depletion of the base region, the depletion voltage of the base region taken from pins 1 and 3 is used as a temperature-dependent measuring parameter, while the external voltage, which shifts the transition in the opposite direction, is applied to pins 1 and 2. By solving the Poisson equation, taking into account the presence of two junctions in the structure (p-n- and p$^+$-n- junctions), we obtain the following expression for the base area depletion voltage:

$$V_0 = \frac{qN_D L^2}{2\varepsilon\varepsilon_0} - V_{K1} - V_{K2} = \frac{qN_D L^2}{2\varepsilon\varepsilon_0} - \frac{2E_G}{q} + \frac{kT}{q} \ln\left(\frac{N_C N_V}{N_D N_{A1}}\right) + \frac{kT}{q} \ln\left(\frac{N_C N_V}{N_D N_{A2}}\right), \quad (3.5)$$

where L is the technological length of the base region; $V_{\kappa 1}$, $V_{\kappa 2}$ is the contact potential difference of p-n- and p$^+$-p junctions of the structure; N_{A1}, N_{A2} are the concentrations of acceptors in p- and p$^+$-areas of the structure, respectively.
As can be seen from equation (3.5), the base area depletion voltage taken from pins 1 and 3 is also a sum of several terms, but depends only on the material parameters of the structure and has no connection with the parameters of the

power supply of the structure, due to which, the measurements are not subject to noise arising in the power supply circuit, and high reproducibility and accuracy of measurement is ensured (Fig. 3.2b), in contrast to traditional diode temperature sensors (Fig. 3.2a).

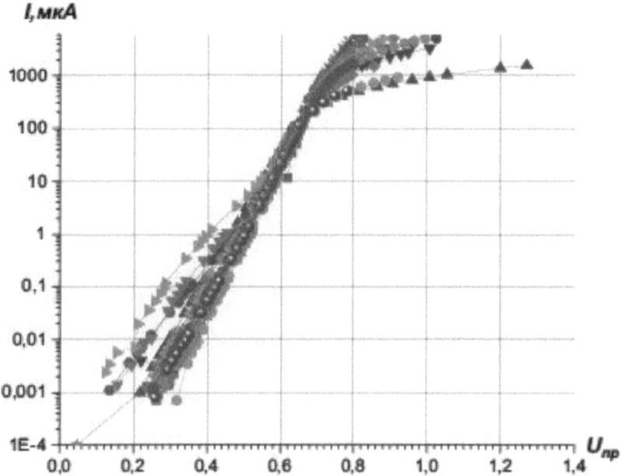

Fig. 3.2b. Volt-ampere characteristics of type II samples.

In addition, according to equation (3.5), the thermosensitivity in the thermosensor with depleted base area has a positive sign, in contrast to traditional thermosensors, in which the sign of the thermosensitivity is negative, which is confirmed by the experimental results presented in Fig. 3.2.

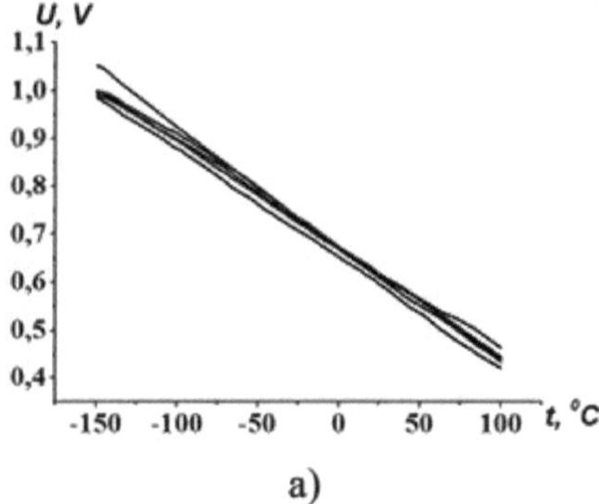

a)

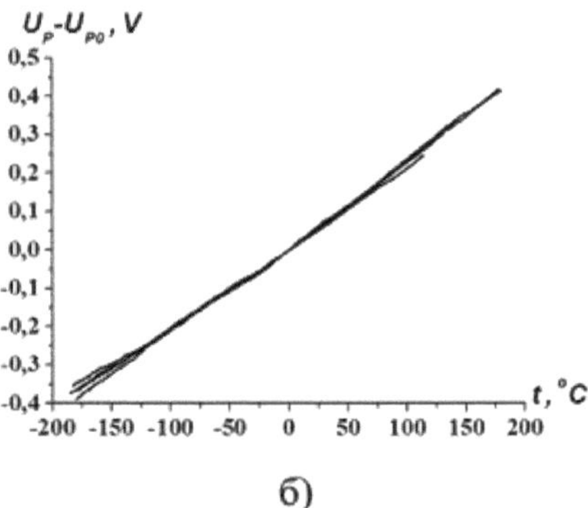

Fig. 3.2. Volt-temperature characteristics of a conventional diode thermal sensor with p-p junction (a) and diode thermal sensor with depleted base area (b)

Thus, analytical expressions (formulas (3.4) and (3.5)) describing the influence of technological and material parameters on the thermosensitivity of the structure for the diode thermosensor with depleted base area and for traditional diode thermosensors are presented, and it is shown that the proposed structure has a significant competitive advantage in improving the accuracy of measurements compared to traditional ones, which is also confirmed by experimental results.

3.3.Investigation of the main performance characteristics of the temperature sensors in the proposed switching scheme

The temperature dependences of the main operational characteristics of laboratory samples of thermal sensors were investigated using the developed and manufactured test bench. In particular, the voltampere characteristics at seven different temperatures were measured (Fig. 3.3.), from which the temperature dependence of the non-ideality coefficient (Fig. 3.4) and saturation current (Fig. 3.5) were determined.

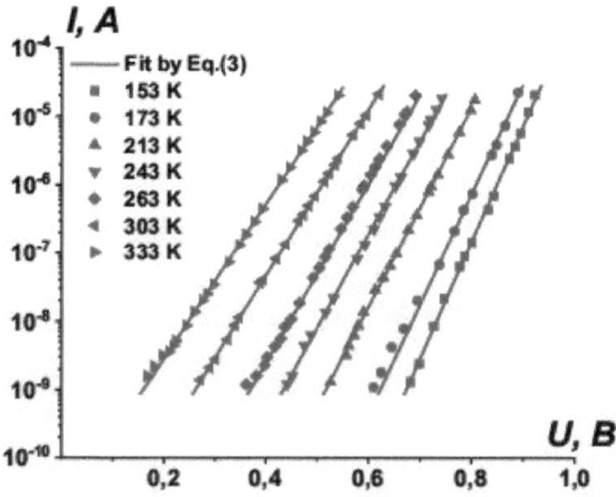

Figure 3.3. Voltampere characteristics at seven different temperatures.
Fig. 3.4. Temperature dependence of the non-ideality coefficient.

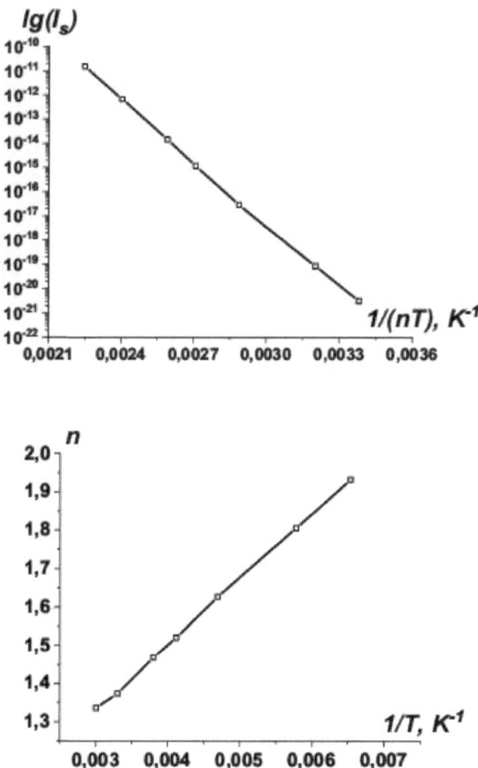

Fig. 3.5. Dependence of non-ideality coefficient and saturation current.

Comparison of the calculated thermal sensitivity values at two different bias currents (1 and 10 µA) with the corresponding experimental values showed good agreement, Fig. 3.6. The temperature dependence (at four temperature values) of the base region depletion voltage on the bias voltage was also measured in Type I (Fig. 3.7) and Type II (Fig. 3.8) laboratory samples. The temperature dependence of the base region depletion voltage at decreasing temperature and at increasing temperature were investigated (Fig. 3.9), which showed the independence (changes are within the measurement error) of the temperature sensitivity from the sign of the temperature change.

Comparison of the calculated thermal sensitivity values (calculated using formula (3.5)) with the corresponding experimental values showed good agreement, Fig. 3.10.

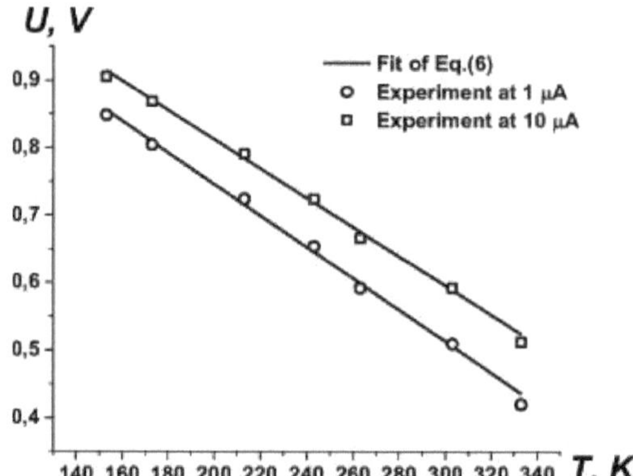

Figure 3.6. Thermal sensitivities at two different bias currents (1 and 10 µA)

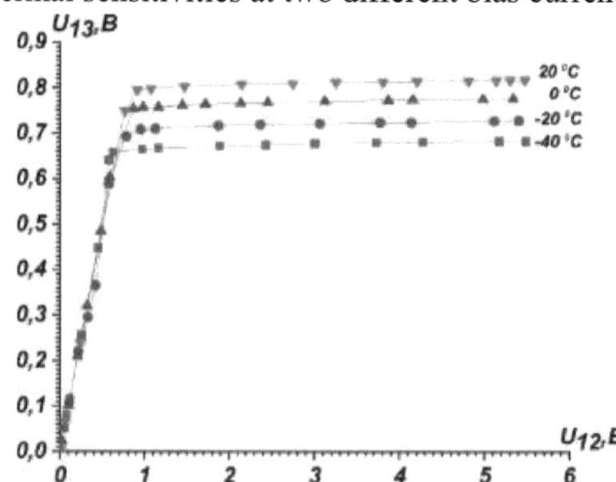

Figure 3.7. Temperature dependences of the base region depletion voltage on the bias voltage in laboratory samples of type I.

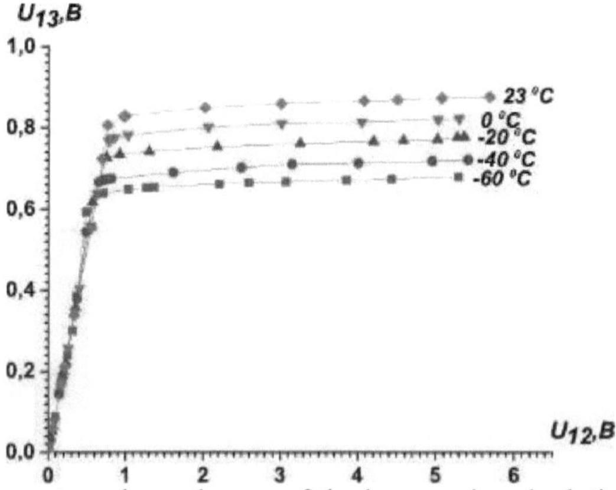

Figure 3.8. Temperature dependences of the base region depletion voltage on the bias voltage in type II laboratory samples.

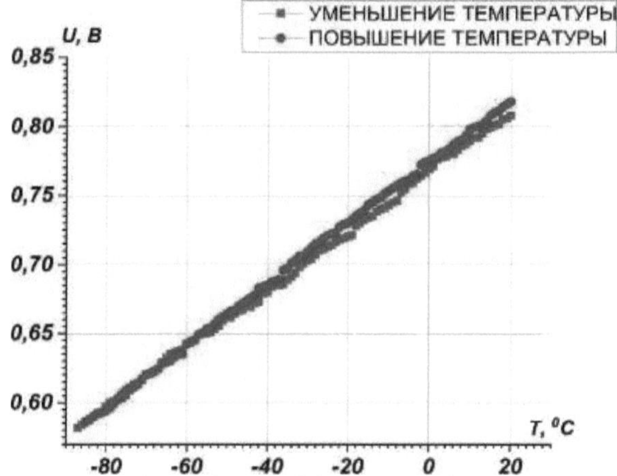

Figure 3.9. Temperature dependence of the base region depletion voltage at decreasing temperature and at increasing temperature.

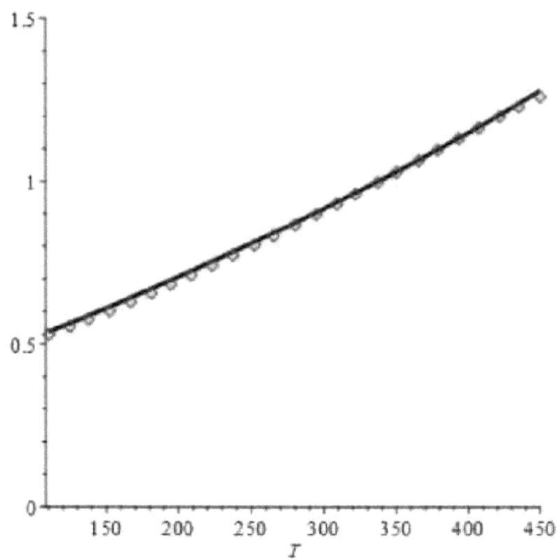

Fig. 3.10. Temperature coefficient of voltage $(TKH = \dfrac{dU}{dT}; \text{мВ/°C})$.

In recent years, temperature sensors have become very widely used in various fields and the demand for them continues to grow every day. In series production it is essential to produce temperature sensors with reproducible and stable parameters, which allows to reduce the costs of rejection and calibration of individual temperature sensors, reduce the number of unusable devices and thus significantly reduce production costs. In this aspect, semiconductor temperature sensors with p-n junction are the most suitable of all devices for use as a temperature sensor, which can be efficiently manufactured using mass production technology, and can be directly integrated with the signal processing and transmission system in a single semiconductor crystal. The best foreign samples of semiconductor integrated temperature sensors today can provide temperature measurements in the range of 0°C to +125°C with an error of ±2.0°C. However, there is a need in science and industry to monitor and measure higher and lower temperatures. Thus, for example, in the steam heating system of urban and rural objects, highly accurate (not worse than ±1°C) measurements of water vapour temperature of the order of (100-150) °C are required.

This thesis work is devoted to the study of the peculiarities of manufacturing of a   three-electrode temperature sensor   and some
technological issues, developed in factory conditions on the basis of silicon structures. It   also reveals the   structural and
thermoelectric features of a three-electrode sensor as opposed to a two-electrode sensor, allowing us to establish the advantages of one or the other structure. In order to create a temperature sensor devoid of the above disadvantages, we have proposed a new structure and design of the thermoelectric sensor. It is a p-1 structure with two contact pads at the top of the n-region of one (n-type), formed and 56

contacting a heavily doped region of a different (p-type) conductivity. The operating voltage is the reverse voltage applied to the contact at the top of the n-region and to the contact (created) in the heavily doped p-region. The measurement potential is taken from one of the upper contact, (in the n-region) relative to the contact (in the lower) heavily doped p-region. At a certain locking operating voltage (depleting p-l-junction) there is a complete depletion on the side of the lower level of the base, and the measuring potential (voltage) from the increasing operating voltage stops increasing and acquires an unchanged value, but with increasing temperature will linearly increase in proportion to it (here the ratio of the increment of the measuring potential to the increase in temperature is "temperature coefficient of thermosensitivity"). As studies have shown, this temperature sensitivity coefficient does not depend on the operating voltage or current, i.e. it has self-stabilising property.

The validity of the obtained experimental results is substantiated by the use in the thesis work of standard laboratory benches, such as L2-28 low-power transistor junction capacitance meter, Shch-300 and 301-1 combined digital instrument, B5-43A constant current source, also 2TRM1 meter-regulator.

Thus, in the thesis a variety of three-electrode structures operating in the thermosensor mode has been experimentally studied and the experimental results obtained show the validity of the calculations and measurements made about the possibility of such a thermosensor operating in the high accuracy mode, with low power consumption and stability when operating from an unstable source with low voltage.

LIST OF REFERENCES

1. Childs P.R.N., Greenwood J.R., Long C.A. Review of temperature measurement. Review of Scientific Instruments, 2000, vol. 71, pp. 2959-2978.

2. Mansoor M., Haneef I., Akhtar S., De Luca A., Udrea F. Silicon diode temperature sensors-A review of applications. Sensors and Actuators A: Physical, 2015, vol. 232, pp. 63-74.

3. Bakhadyrkhanov M.K., Valiev S.A., Tachilin S.A., Nasriddinov S.S. Sensitive thermosensors on the basis of highly compensated silicon. Surface Engineering and Applied Electrochemistry, 2007, vol. 43, no. 6, pp. 505-507.

4. Karimov M., Makhkamov Sh., Makhmudov Sh.A., Muminov R.A., Rakhmatov A.Z., Sandler L.S., Sattiev A.R., Sulaimanov A.A., Tursunov N.A. Peculiarities of influence of radiation defects on photoconductivity of silicon irradiated by fast neutrons. Applied Solar Energy, 2010, vol. 46, no. 4, pp. 298-300.

5. Udrea F., Santra S., Gardner J.W.. CMOS temperature sensors - concepts, state-of-the-art and prospects / CAS 2008, International Semiconductor Conference, Sinaia, Romania, 2008, 13-15 October. -PP. 31-40.

6. Souri K., Chae Y., Makinwa K.A.A. A CMOS Temperature Sensor With a Voltage-Calibrated Inaccuracy of ±0.15°C (3σ) from -55°C to 125°C // IEEE Journal of Solid-State Circuits, 2013. - Vol. 48, No. 1. -PP. 292-301.

7. Patent RUz № IAP 05120 "Multifunctional sensor based on field-effect transistor" / Karimov A.V., Yodgorova D.M., Abdulkhaev O.A., Dzhuraev D.R., Turaev A.A..

8. Physics. Big Encyclopaedic Dictionary / Ed. by A. M. Prokhorov. Moscow: Big Russian Encyclopaedia, 1998. C. 741- 944.

9. Evdokimov I.N. Methods and means of research Part 1. Temperature (textbook). Gubkin Russian State University of Oil and Gas, Department of Physics, Moscow: 2004, 106 p.

10. Units of physical quantities and their dimensions: Training and reference manual. Moscow: "Nauka", 1988. 432 c.

11. Д. А. Parshin, G. G. Zegrya. Statistical thermodynamics.

12. GOST 8.417-2002. State system for ensuring uniformity of measurements. UNITS OF MAGNITUDES.

13. Galileo G. Assay master - M., 1987.

14. Bulkin P.S., Vasilieva O.N., Kirov S.A., Malova T.I. Temperature measurement by semiconductor thermometers. Tutorial - M.: OOP Phys. facta MSU, 2012, 3-5s.

15. Gerashchenko O. A. Thermal and temperature measurements. Reference manual. K.: Nakova Dumka, 1965, 304 p.

16. GOST 27544-87. Liquid glass thermometers.

17. Grunin V. K. §2.3.4 Thermoelectric radiation receivers// Radiation sources and receivers:-SPb.: Publishing house of SPbGETU "LETI", 2015. - 167 c.

18. V.I. Panferov. To the theory of thermocouples. - Bulletin of SUSU, 2007,№14,48-50 p.

19. Zhukovsky, P. A. Review of metallic thermoresistive sensors / P. A. Zhukovsky, M. K. Avseenok. - Text : direct // Young scientist. - 2019. - № 28 (266). -C. 30-31. https://moluch.ru/archive/266/61588.

20. Abdulhaev O.A., Bebitov R.R., Abdulhaeva A.R., Yodgorova D.M. Limitation of accuracy of temperature measurement by semiconductor sensors depending on the stabilisation coefficient of the operating current // Uzbek Journal of Physics, 2018. - Vol.20, No.5. - PP. 300-304.

21. Abdulkhaev O.A., Yodgorova D.M., Bebitov R.R., Hakimov A.A., Rakhmatov A.Z., Shertoev J.Kh Features of thermal sensitivity of silicon structures with depleted base area // "Physics of semiconductors and microelectronics", 2019, Volume 1, Issue 3, pp.43-50.

22. Zaitsev, Yu.V. Semiconductor thermoelectric converters // M.: Nauka, 1985. 120 c.

23. Qualification work SUSU -12.03.01.2017.114 vkr.- P.35.

24. Raymond G., Temperature sensors: contact or noncontact? // Sensors magazine, Jan2006. http://www.sensorsmag.com/sensors/article/articleDetail.jsp?id=317360

25. Gordov A.N. Fundamentals of pyrometry // 2nd ed. M.: Metallurgy, 1971.

26. Ambrok G.S., Baronenkova Y.D., Gogolev H.JI. Methods and Means of Optical Pyrometry // M.: Nauka, 1983. C. 93-102.

27. Samsonov G.V., Kitz A.I., Kyuzdeni O.A. Sensors for Temperature Measurement in Industry // Kiev: Izd-wo nauk, dumka, 1972 -251s.

28. Mathews D. Choosing and using a temperature sensor // Sensors magazine,121 https://mirmarine.net/elektromekhanik/sudovaya-avtomatika/979-datchiki-temperatury

29. Golembo B.A., Kotlyarov B.JT., Shvetsky B.I. Piezoquartz analogue-digital temperature converters // Lviv: Vishcha Shkola, 1977.

30. Physical quantities. Reference book. Moscow: Energoatomizdat, 1991. http://twt.mpei.ac.ru/PVHB/index.html

31. A. J. Chiquito, O. M. Berengue, E. Diagonel, J. C. Galzerani, and J. R. Moro, "Temperature sensors based on synthetic diamond films," Diam. Galzerani, and J. R. Moro, "Temperature sensors based on synthetic diamond films," *Diam. Relat. Mater.*, vol. 16, no. 8, pp. 1652-1655, 2007, doi:

10.1016/j.diamond.2007.02.012.

32. D. L. Blackburn, "Temperature measurements of semiconductor devices - A review," in *Annual IEEE Semiconductor Thermal Measurement and Management Symposium*, 2004, vol. 20, pp. 70-80, doi: 10.1109/stherm.2004.1291304.

Printed by Books on Demand GmbH, Norderstedt / Germany